IMAGES
of America

Portsmouth Harbor's Military and Naval Heritage

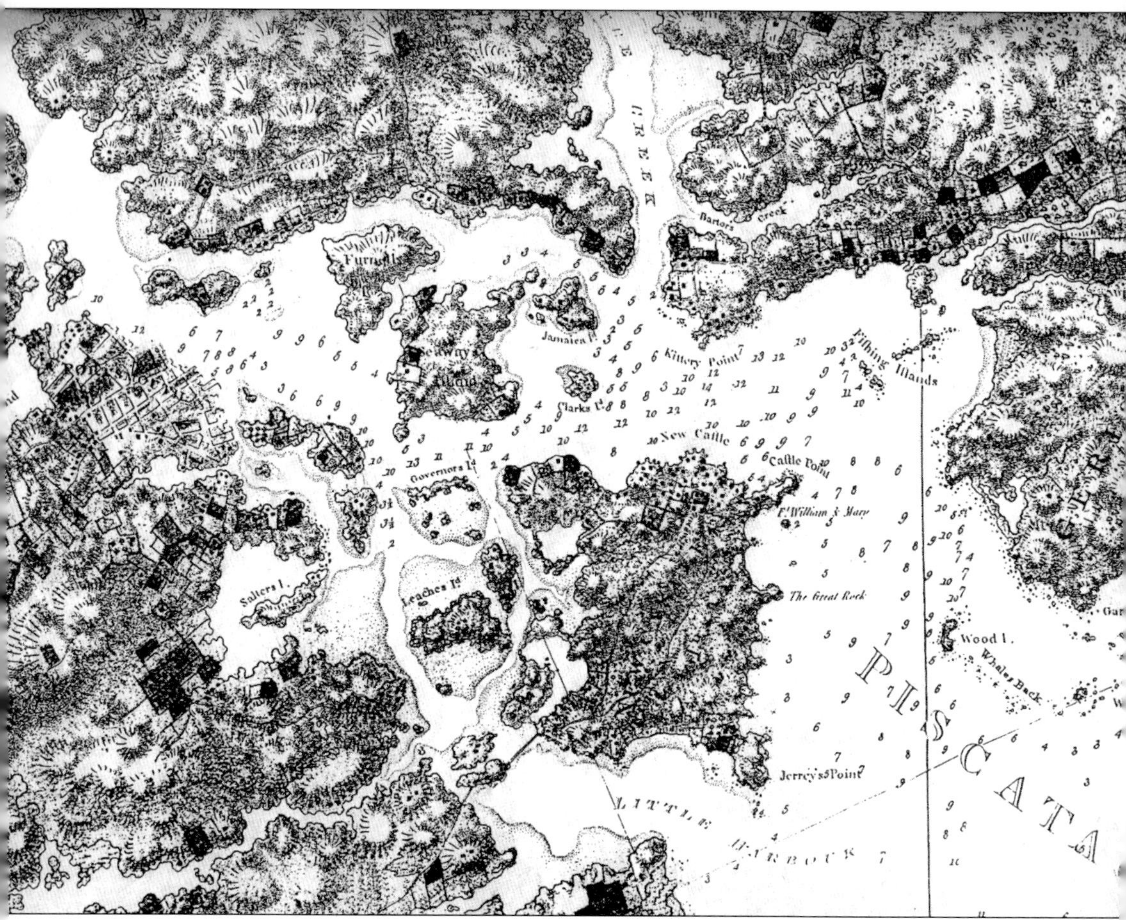

This 18th-century map of the lower Piscataqua River reveals Furnalds (later Dennetts) Island and Seaway (later Seaveys) Island in Kittery, Maine, directly across the river from Portsmouth, New Hampshire. During the next century, these islands were developed into the Portsmouth Navy Yard. On Great Island, at the mouth of the river, Fort William & Mary stands on Castle (or Fort) Point. Later, harbor defense batteries or forts were established at Jerrys Point in New Castle, New Hampshire, on Gerrish Island in Kittery Point, Maine, and at Frost and Odiornes Points in Rye, New Hampshire. (NH.)

IMAGES
of America

Portsmouth Harbor's Military and Naval Heritage

Nelson H. Lawry, Glen M. Williford, and Leo K. Polaski

ISBN 0-7385-3647-4

First published 2004

Published by Arcadia Publishing,
Charleston SC, Chicago IL, Portsmouth NH, San Francisco CA

Printed in Great Britain

Library of Congress Catalog Card Number: 2004107270

For all general information, contact Arcadia Publishing:
Telephone 843-853-2070
Fax 843-853-0044
E-mail sales@arcadiapublishing.com
For customer service and orders:
Toll-free 1-888-313-2665

Visit us on the Internet at www.arcadiapublishing.com

On the cover: A briskly rowed cutter flies the national colors in line with USS *Jamestown*. Astern of her in this 1889 photograph is the Portsmouth Navy Yard's receiving vessel, the old frigate *Constitution*. (PS.)

In 1897, upon the order of the secretary of the navy, the old *Constitution*, long a receiving ship at the Portsmouth Navy Yard, was taken to the Boston Navy Yard, and in 1905, she was partially restored. Fully returned to her previous dignity a quarter-century later, the newly recommissioned USS *Constitution* appears here during her July 1931 visit to the Portsmouth shipyard. The 44-gun sail frigate has undergone another major restoration since, for her 200th birthday in 1997, and remains the oldest commissioned ship in the U.S. Navy. (SB.)

Contents

ACKNOWLEDGMENTS

This book could never have been written without the generous assistance of individuals at the Portsmouth area's major historical archives who opened their collections to us, vastly added to our knowledge of military, naval, and civilian activities, and encouraged our undertaking at every step. Tara Webber, Strawbery Banke Museum's registrar and librarian, along with James Dolph and Walter Ross of the Portsmouth Naval Shipyard Museum, were unfailingly supportive, and both collections produced photographic treasures we are still marveling about. Wayne Manson, director of the Kittery Historical and Naval Museum, added greatly to our information about the shipyard and naval activities. Rebecca Ernest and Nancy Mason, Special Collections librarians at the University of New Hampshire's Dimond Library, welcomed us every time. Pam Cullen of the New Castle Town Hall kindly provided repeated access to the town's historical photograph collection. Nicole Cloutier, Special Collections librarian at the Portsmouth Public Library, and Catherine Beaudoin, director of the Dover Public Library, were generous with their time and resources. History writer Richard E. Winslow III collegially researched two difficult questions and shared his findings with us.

We wish to recognize the generosity of the following individuals, working from more distant locations: Bolling Smith, for sharing the relevant photographs from his collection of period images; Terrence C. McGovern Jr., for assisting an aerial photography trip; National Park Service writer and historian Carole Perrault, for contributing photographic and textual material she had acquired for a Fort Constitution project; and Dr. Joel W. Eastman and Gregory J. Hagge, for resolving a particularly troublesome point in our research. Additional photographs and information most useful to the book were found at the National Archives and the Library of Congress. Finally, thanks beyond measure are due to Wynn S. Shugarts, who ably served as recorder, researcher, wordsmith, photo handler, and supporter.

Photographs are keyed as to source by the following notations: Bolling Smith Collection (BS); Corps of Engineers Historical Section (EM); Glen Williford Collection (GW); Kittery Historical and Naval Museum (KM); Library of Congress (LC); National Archives, Washington/College Park (NA); New Castle Historical Collection (NC); University of New Hampshire (NH); National Oceanic and Atmospheric Administration (NO); National Archives, Waltham Regional Branch (NW); Portsmouth Public Library (PL); Portsmouth Naval Shipyard Museum (PS); and Strawbery Banke Museum (SB).

The new 22nd Coast Artillery Regiment was activated in 1940 and eventually was filled out to man the entire fixed gun and mine defenses of Portsmouth. The regiment was headquartered initially at Fort Constitution, though within a few months that function was transferred to Camp Langdon. The unit crest shows an early ship, perhaps *Bedford Galley*, surrounded by a red background and riding on blue and white waves, reminiscent of the New Hampshire state seal. (GW.)

INTRODUCTION

The major source of the Piscataqua River is the south- and east-running Salmon Falls (or Newichawannock) River, which centuries after the first European habitation, still serves to demarcate New Hampshire from Maine. Flowing southeastward, the Piscataqua makes a final, southward bend past Great Island into the Gulf of Maine. The lower part of the Piscataqua estuary constitutes Portsmouth Harbor, serving the city of Portsmouth and the town of New Castle, New Hampshire, on its south bank, and the town of Kittery, Maine, on its north bank.

Historically, the sizable harbor at the mouth of the Piscataqua River has been an important one, first attracting interest from British and French sailor-adventurers in the early 17th century. Because of its proximity to the almost endless virgin forest of immense white pine, Portsmouth served as the Royal Navy's most important—and for much of the period, its only—North American mast port during several decades overlapping the 17th and 18th centuries.

During the interval from 1630 to 1633, the colonists erected a small earthwork, with a "fort-house" and four brass cannon, on the northeastern projection of Great Island, which was soon known as Fort Point. This first defensive work was likely built as protection against pirates, as that period coincided with the depredations by Dixy Bull's marauding band.

The construction of a more substantial timber-palisaded work owed to the series of naval wars between Great Britain and the Netherlands. It became even more of a formalized, professionally built fortification during the early 1690s, and was the namesake of the reigning monarchs, William and Mary. Until the end of the 18th century, Fort William & Mary—and after 1720, its counterpart on the Maine side, originally named Fort William—protected the upriver shipyards, which built 10 warships for four different navies.

Fort William & Mary gained attention in 1774, when American patriots raided it and removed both gunpowder and small cannon. The original site remained the same for successive United States forts of the First, Second, and Third Systems, rebuilt until the end of the Civil War. Early in the 19th century, the names of the two harbor defense works became Fort Constitution (at New Castle) and Fort McClary (at Kittery Point), and the former saw a Martello tower, named Walbach Tower, erected to protect its rear.

In 1800, the new shipyard for the U.S. Navy was established on Dennetts (or Fernalds) Island on the Kittery side of the river. Capt. Isaac Hull assumed his duties in 1813 as the Portsmouth Navy Yard's first commandant. During his tenure, the yard constructed the first of three great shiphouses, where sailing men-of-war were initially built, and within decades, their steam-propelled successors. Following the Civil War, the funding for both services diminished to abysmal levels, and amidst talk that the yard would be sold, the garrisons of the harbor forts were replaced by caretaker sergeants.

Momentous happenings, however, were about to touch both the military and naval branches at Portsmouth. Although delayed for a decade after the Endicott Board issued its report in 1886, the construction of eight modern batteries, of five different calibers, took place at three harbor defense works: Fort Constitution at Fort Point and Fort Stark at Jerrys Point, both in New Castle, and Fort Foster on Gerrish Island in Kittery Point. The larger guns were mounted on the ingenious, complicated, and ultimately range-limiting disappearing carriage. Additionally, the provision was made for planting fields of powerful submarine mines, including the construction of shoreside control and storage facilities.

Even more affected by the Spanish-American War, the Portsmouth Navy Yard saw, in very close progression, the coming and going of Spanish naval prisoners of war and their marine guards; the conversion of Jenkins Gut between closely adjacent Dennetts and Seaveys Islands into a capacious dry dock; the construction of an enormous cofferdam and the explosive destruction of hazardous Hendersons Point; the arrival and departure of the envoys meeting to negotiate the conclusion of the Russo-Japanese War; and the construction, on the site of the Spanish camp, of the Portsmouth Naval Prison, soon to be known as "the Castle" from its distinctive architecture.

The yard's first submarine, *L-8*, was launched in August 1917, while America was at war. For the first time ever, women—both civil servants and naval personnel—arrived at the shipyard to perform a multitude of jobs. Wartime construction of submarines spilled over into the 1920s, and such building and testing continued during the interwar years.

The coast defenses were garrisoned during World War I, but unlike such defenses elsewhere, no new batteries were constructed to protect Portsmouth Harbor. Development of the New Reservation was undertaken at New Castle, years later to be renamed Camp Langdon and become the harbor headquarters post. Between the wars, the harbor defenses again were reduced to only caretaker status. But as the world situation became worse, in 1940, the 22nd Coast Artillery Regiment was formed to man Portsmouth's defenses.

Disasters impacted the yard in the years before the onset of World War II, including the sinking of one of its boats, USS *Squalus*, with the drowning of nearly half her crew. More men were saved than lost, and *Squalus*, recovered, restored, and renamed *Sailfish*, went on to compile an excellent war record.

During World War II, the Portsmouth Navy Yard delivered 69 fleet submarines, most of which served in the Pacific and helped decimate the enemy's navy and merchant marine. All the while, the yard was protected by the harbor defenses, whose members initially manned the old Endicott-era gun and mine batteries, as entirely modern ones underwent construction. The largest was a battery of two 16-inch rifles, reaching 25 miles seaward, built at the new military reservation, Fort Dearborn, in Rye, New Hampshire. A series of 14 fire control stations dotted the coast from Massachusetts to Maine, although soon they were relegated to a backup role by the newly developed fire control radar.

In 1944, the war was decidedly winding down, and with the danger to the East Coast almost negligible, the 22nd Coast Artillery underwent successive reductions in force. With the number of enemy vessels in the Pacific becoming increasingly scarce, submarine construction at the yard also declined. The postwar years saw the modernization of existing fleet boats, and then the building of entirely new and radical underwater craft: those having nuclear power. But in a world of fission and fusion bombs delivered by ballistic missiles, harbor defense by fixed gun batteries was gone forever, and Pease Air Force Base materialized to confront that terrible new threat.

Two buddies at Fort Foster in World War II hang onto a Lister bag, which was usually set up in the field for soldiers to fill their canteens from. Here, the Lister bag has been erected near the soldiers' barracks, probably to give them the delightful experience of sampling the iodine-purified water they will encounter during training and in combat areas. The water was safe to drink but always tasted a bit like canvas. (KM.)

One

At Short Range
The Muzzleloading Era

The fine natural harbor created at the mouth of the Piscataqua River was attractive for early settlement and commercial development, including the mast trade. The British colonists began fortifications as early as 1631. This first work and a stronger one 30 years later were located on the New Hampshire side at Great Island (the present New Castle Island). After additional stores and cannon were shipped there, a much-enlarged work of the 1690s was named Fort William & Mary, after the reigning monarchs. By 1721, this site was complemented by a work on the Maine shore at Kittery Point, named Fort William.

During times of threat, other sites were also developed for temporary defensive fortifications. These were most likely simple, militia-manned earthworks and could be found at Jerrys Point, otherwise called Jaffreys Point, and on some of the upriver islands, such as Fort Sullivan on Seaveys Island and Fort Washington on Pierces Island, built during the Revolutionary War. A few of these sites were periodically repaired and then utilized in the War of 1812.

American defenses commenced when the state of New Hampshire transferred the site of Fort William & Mary to the federal government in 1791. The work there, renamed Fort Constitution in 1801, was an important American coast defense position until after World War II. In 1808, the fortified location on the Kittery side was obtained; soon named Fort McClary, it became one of the essential defenses of Portsmouth. While rebuilt and modified several times, the forts had both a similar general structure and armament. They consisted mostly of a raised earthen rampart, faced with either stone or brick masonry. The ordnance consisted of heavy, smoothbore, muzzleloading guns made of iron, typically firing a 24-pound or 32-pound round shot. Fort Constitution usually had 20 to 40 such weapons and smaller Fort McClary, fewer than a dozen.

At certain times throughout this period, the harbor forts were garrisoned and saw active defense, such as during the War of 1812, while at others, just a single caretaker minded the gunpowder store.

The most frequently fortified site in Portsmouth Harbor is Fort Point on Great Island, now known as New Castle Island. Though at times in a state of virtual abandonment, this location possessed important defenses almost continuously from 1631 to 1950. The point projects into the channel of the Piscataqua River, where the river changes course, forcing vessels to slow for the turn. (GW.)

The first record of a fortification at Fort Point dates from 1631 and reveals nothing more than a simple earthen redoubt or fort house. Substantial rebuilding occurred in the mid-1660s and the early 1690s, when the fort was named Castle William & Mary, after the British monarchs who funded its extensive improvements. The notches in the wall of the work, as seen from the Piscataqua River, are embrasures, one for each of the approximately 60 cannon of the intended armament. (KM.)

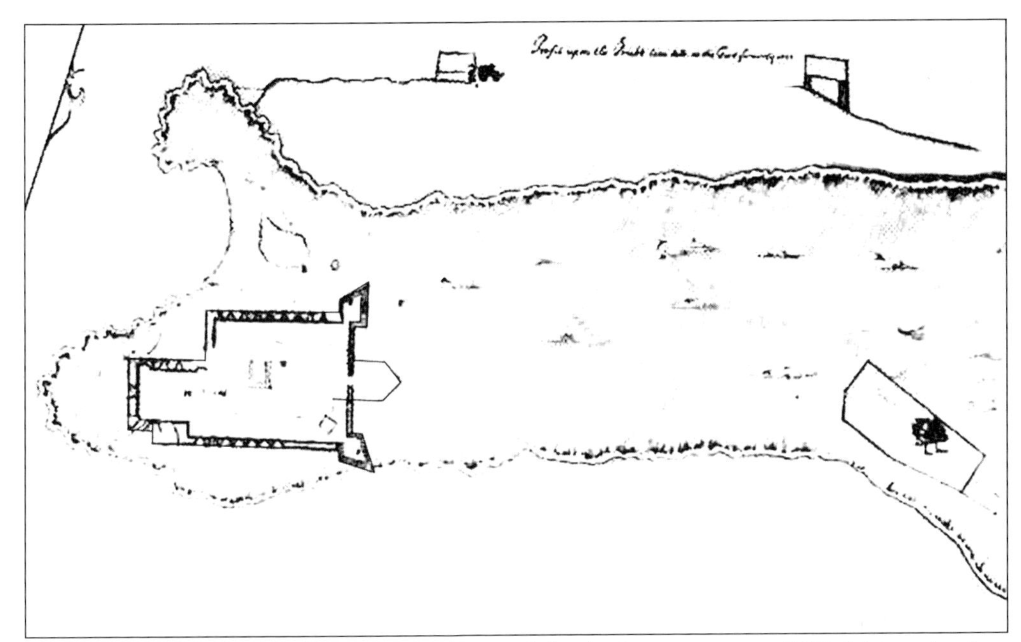

The first masonry walls were constructed around the fort's perimeter in 1705. This early plan, originally from the collection of the New Hampshire Historical Society, reveals the fort's outline as of that date. Projecting bastions gave additional protection to the landward approaches across the narrow neck of land. (KM.)

Royal Governor John Wentworth presided over the province of New Hampshire from 1767 to 1775. In mid-December 1774, two separate parties of local patriots, under the command of John Langdon and John Sullivan, overcame the garrison at Fort William & Mary and carried off the fort's gunpowder supply, which was desperately needed by the revolutionaries. These raids were among the first organized acts by the rebels, and as such, firmly connect the fort on Great Island to American history. (LC.)

The state of New Hampshire transferred the land containing Fort William & Mary to the federal government in 1791. The fort was repaired and rearmed between 1794 and 1798, during what is known as the First System of American fortifications. While no formal naming authority has ever been located, Secretary of War Henry Dearborn first referred to the defenses as Fort Constitution in 1801. The fort was again rebuilt, its scarp raised and capped with stone, in 1808. (LC.)

The general form of the fortification at Fort Point followed approximately the same trace for almost 150 years. The fort had roughly a rectangular form, with three linear sides, and a seafront northeastern extremity of semicircular shape. Most guns were placed atop the ramparts on this circular front, with others firing mainly to the north. The main gate was on the western front. A heavy wooden door was set in a masonry entry hall, with guardrooms on either side. This drawing likely dates between 1790 and 1820. (KM.)

Though the main entry into the fort was rebuilt several times, most recently in 1974 for the bicentennial celebration, it very much resembles the original one. This picturesque entryway, with its contrasting stone and brick, was always a favorite spot for snapshots. Here, a group of soldiers prop up the portcullis with timbers *c.* 1900. Buildings have been placed close to the fort's walls because of the limited land available on the neck, and a temporary water pipe has been run along the ground. (BS.)

Col. John de Barth Walbach commanded Fort Constitution from 1806 to 1821. An Alsatian by birth, he immigrated to the United States and served the army in a variety of posts. An outstanding officer, Walbach also made a favorable impression on Portsmouth—he took responsibility for the fatalities caused by a premature gunpowder explosion during an 1809 Independence Day celebration, even though he was entirely blameless. Walbach did not retire from the army until the age of 93. (LC.)

In 1814, after enjoying freedom from the British naval blockade imposed on the lower Atlantic Coast during the War of 1812, New England was finally blockaded. A defensive tower, later named Walbach Tower, was built on Jourdans Rock to protect the rear of Fort Constitution and command New Castle's Town Beach. A 32-pounder gun was placed atop the tower, and marksmen and small guns fired from embrasured casemates below. (KM.)

Walbach Tower was constructed very quickly in September 1814. Later reports even claim that it was built by volunteers in a single night. While this is an exaggeration, the hurried construction has not endured the passage of time well. The stone and brick have collapsed, so that today few recognizable features remain. In this World War I view, however, one of the lower embrasures of a 4-pounder gun is still in relatively good shape. (NC.)

The plan for rebuilding Fort Constitution in the 1840s proved more a modernization of the existing 1808 work than new construction. The outer scarp walls were raised, and a stone coping was placed on the walls of the parade and ramp. The old wooden platforms for the guns were replaced with permanent stone ones. The 1842 plan shows emplacements for 41 guns on the sea-front ramparts and a single gun in the south bastion. The large building interior to the semicircular wall is the citadel, intended for a last stand. (NA.)

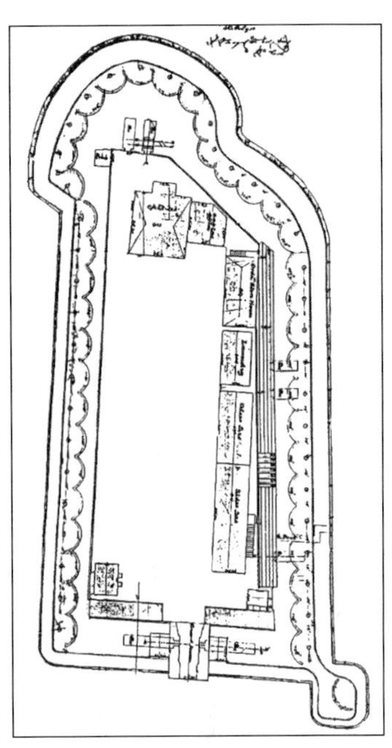

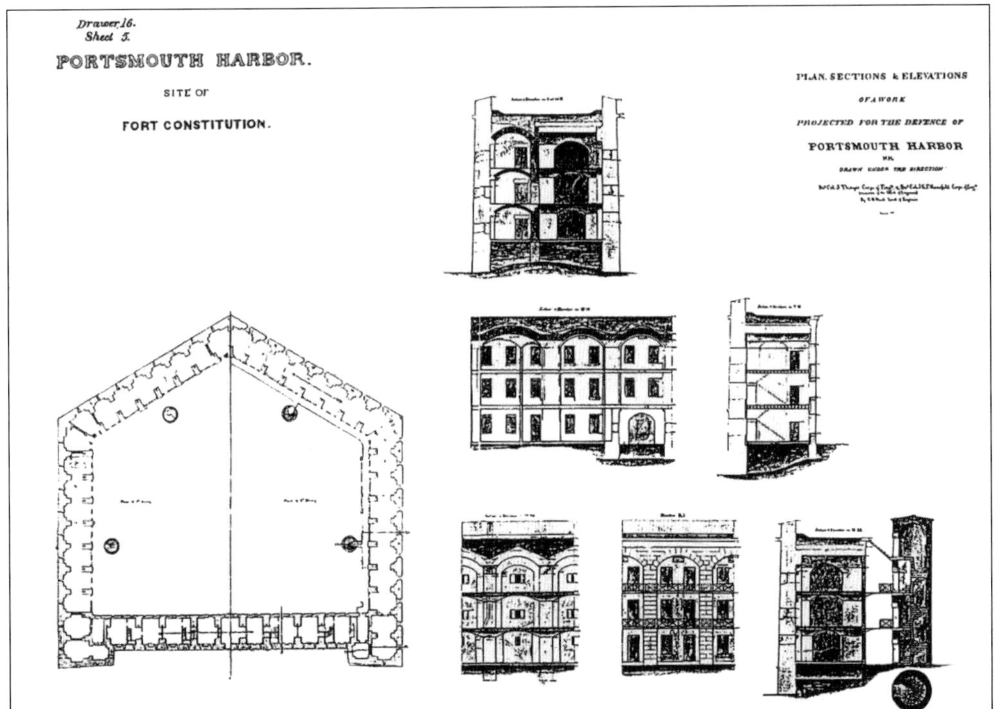

In the early 1860s, approval was granted to construct a completely new fort at Fort Point. This 1861 shows the design, essentially a truncated hexagon. It was intended to have three layers of casemated guns and a layer of guns on the topmost barbette level, and carry a total of 149 guns. Small demibastions were to project from the landward face and provide flanking-fire protection for the main gate. After the end of the Civil War, construction abruptly ceased. (NA.)

In 1771, the British constructed a wooden lighthouse at Fort William & Mary in New Castle, now the site of Fort Constitution. It was replaced in 1804 by an 80-foot-high octagonal timber structure, shortened to 55 feet in 1854. Marking the entrance to Portsmouth Harbor, this tower warranted a fourth-order Fresnel lens—a 28-inch, hand-polished prismatic glass lens, shaped like a beehive, which focused light from an oil lamp into horizontal beams to increase its visibility from the sea. (NC.)

Because of the danger of fire, the wooden lighthouse was replaced in 1879 with a 48-foot-high cast-iron tower, built on the same foundation and lined with brick. Inside, a cast-iron spiral staircase led to the lamp house. Its fourth-order lens was electrified in 1934 and was fitted with a cylindrical green filter to distinguish this navigational aid from others. The ashlar blocks intended for the 1860s construction and the timber falsework for erecting the casemate arches are evident. (KM.)

This 1886 view reveals a tranquil setting at the Fort Constitution reservation. The post's wharf is on the left, and the southwest-facing gorge wall is in the right center. Work began on the new fort on November 16, 1862. Local contractor William Mathes provided most of the labor, though an engineer officer was always present. Only the left-most wall is new construction. Priority was given to the cannon-armed casemate fronts, and work never progressed to the rear gorge wall. (KM.)

Eventually, the new construction completely surrounded the site of the 1808 Fort Constitution. Some of the older work was retained in order to provide defense during construction but was to be entirely removed upon completion of the new fort. When construction was suspended after the Civil War, however, a fort-within-a-fort effect occurred, as clearly seen in this 2002 aerial photograph. (GW.)

The bluff on which Fort McClary was situated lay only 1,200 yards north-northwest of Fort Constitution. Although the deep part of the Piscataqua River ran closer to Fort Point, the position still proved useful in providing crossfire on sailing vessels negotiating the bend in the channel. In fact, the site had first been fortified *c.* 1720 by Massachusetts to reinforce its negotiating stance with New Hampshire in the matter of duties charged on merchant traffic. (KM.)

The distinctive blockhouse, an anachronism even at the time it was built in the mid-1840s, is the most prominent feature of Fort McClary. The structure to the right in this 1950s postcard view is one of two brick riflemen's houses, with narrow vertical firing slits. Behind the blockhouse stands the early powder magazine. Late in the 19th century, the navy stored gunpowder from ships being repaired at the Portsmouth Navy Yard in the fort's 1863 magazine. (GW.)

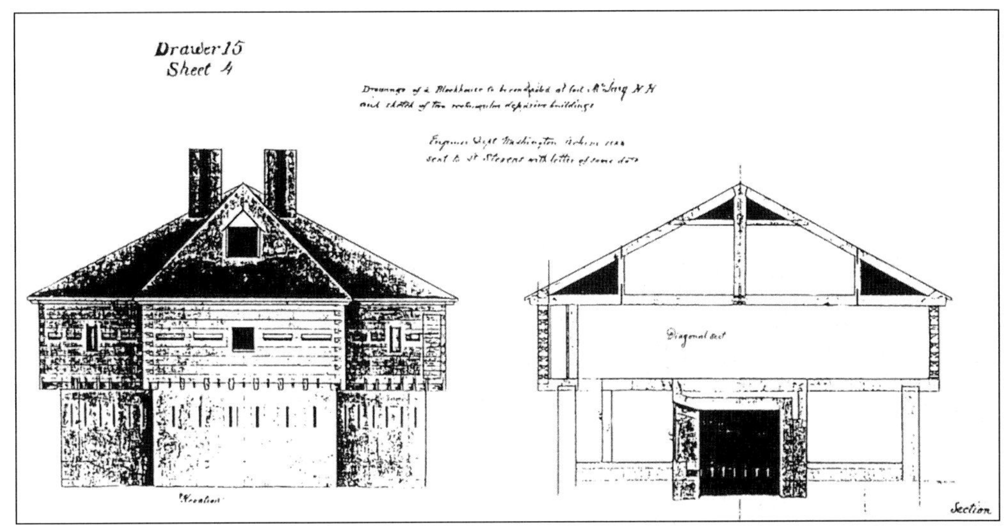

In the early 1840s, the U.S. Army received approval to make major repairs and enhancements to older forts. Both Forts Constitution and McClary benefited from this initiative. At that time, Fort McClary received its substantial blockhouse, shown in this 1844 plan. Interestingly, the construction was a combination of granite stone on the lower level and heavy timbers on the upper. Firing slits for muskets, as well as a few openings for light guns, were incorporated into the design. (NA.)

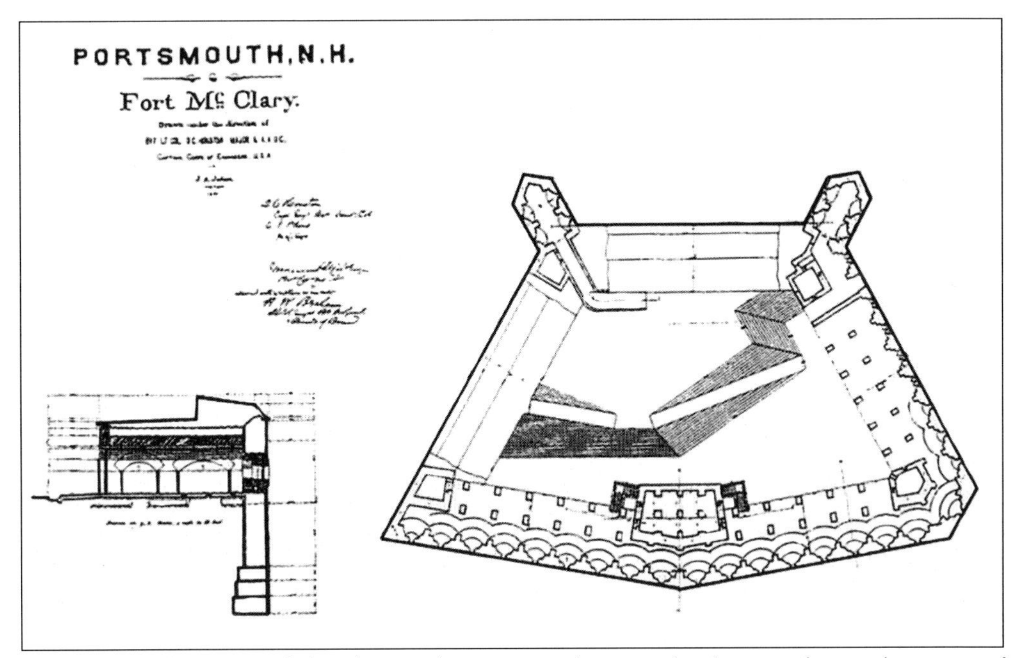

Congress finally approved the plans and appropriated money for the complete replacement of old Fort McClary just as the Civil War erupted in 1861. This plan, dated that year, reveals a substantial casemated work, with a heavy sea-front gun battery and an enclosed parade ground with flanking bastions for landward approaches. Built of fine New England granite, it would have been an imposing work capable of providing a strong defense. (NA.)

During the Civil War, the destructive effect upon vertical masonry walls by rifled guns firing explosive shells became dramatically apparent. At virtually all coastal forts still under construction, work ceased permanently within a few years after the war's end in 1865. Only the initial construction had been accomplished on Fort McClary's outer walls, and it was never completed. The condition of the walls today is identical to this *c.* 1905 view. (SB.)

This aerial photograph includes the major elements of Fort McClary's three major fortification plans. The cluster of buildings on the central rise includes the 1840s blockhouse and riflemen's house and the 1808 powder magazine. The outer granite wall, roughly a pentagon with a projecting caponier, belongs to the 1860s construction program. The earthen rampart of a small, circular-shaped gun battery, just visible to the left of the blockhouse, appears on the trace of the 1808 Second System battery. (GW.)

During the Civil War, the North feared that Confederate warships would attempt to raid the New England coastal trade, or perhaps even attempt a quick attack on a port or naval dockyard. Newly manufactured guns were thus distributed to many older forts and a number of emergency earthworks. In this *c.* 1900 photograph of Fort McClary, several smoothbore Rodman (left) and rifled Parrott guns have been stored near the old water battery below the blockhouse since the war. (KM.)

Because the Spanish-American War found the United States coastline inadequately defended, a substantial number of obsolete muzzleloaders from the Civil War were brought back for emergency use. The army mounted three such Rodman guns, languishing at Fort McClary for decades, in quickly prepared emplacements. Although the guns would have been of questionable effectiveness against a modern fleet, they did make great places to sit and pose. (KM.)

During the 1880s and early 1890s, the harbor defense forts of Portsmouth led a peaceful, often ignored, existence. Only the annual militia practice or a significant celebration broke the lax routine. Here, in 1893, bronze Napoleons—muzzleloaders left over from the Civil War 30 years earlier—are fired for the bicentennial celebration of the town of New Castle, just outside Fort Constitution. The town owes its name to the "new castle" built the year before its 1693 incorporation. (BS.)

Here, civilian picnickers make fine use of one of the characteristics of the dressed granite blocks used to construct parts of Fort Constitution: a smooth finish. Stacked in piles of differing heights, some have become convenient seats and a picnic table, while others, piled higher, allow men to perch. The blocks soon joined their fellows, hoisted by a winch and derrick, to pave the casemates of the unfinished fort. (SB.)

Two

Wood, Sail, and Steam

The Navy Yard, Part I

Portsmouth Harbor served as the Royal Navy's premier North American mast port during part of both the 17th and 18th centuries. In the 1690s, local shipwrights built two small, lesser-rated warships for that navy; during the 18th century, another eight men-of-war, ranging in size from revenue cutter to ship-of-the-line, slid down the ways. Whereas all such construction had been done on Badgers Island, the new building yard established in 1800 for the U.S. Navy was on 58-acre Dennetts Island in Kittery, Maine (then part of Massachusetts). The first marines arrived during 1806 to protect the property, and seven years later, Capt. Isaac Hull assumed his duties as the Portsmouth Navy Yard's first commandant.

By 1838, the yard had three massive shiphouses—Alabama, Santee, and Franklin, named after the most significant vessels built in each. Nine years later, the yard laid down *Saranac*, its first steam frigate, albeit a sidewheeler. At a cost well exceeding $1 million, in 1852, the yard completed a floating dry dock of 5,000 tons displacement, which was used almost immediately to move the venerable frigate *Constitution* onto building ways ashore for renovation. During the subsequent Civil War, the yard built 26 vessels—mostly screw sloops-of-war, ironclad monitors, and sidewheel steamers.

Despite purchasing part of Seaveys Island in 1862 and its entirety four years later, in the postwar years, the shipyard became nearly moribund. A proposal was made in 1876 to sell off the yard, but wiser heads prevailed and the facility was retained. Eight years thereafter, two events of significance occurred. First, the decommissioned sail frigate *Constitution* arrived at Portsmouth and was converted to the yard's receiving ship. Second, the navy mounted an expedition to rescue the survivors of the Greely arctic expedition, purchasing two former sealing vessels to accomplish the job. Lt. Adolphus Greely, U.S. Army, and five survivors were brought to the yard to recover their health.

Although transitional wooden screw warships were still in use, by the late 1880s, the navy had embarked on the construction of modern steel warships mounting breechloading guns. While no new construction took place at the Portsmouth Navy Yard, the facility continued with refit and repair work sufficient to keep the old navy afloat long enough to be replaced by the new. The long-discussed naval hospital on Seaveys Island became a reality in 1891 and would soon see use, as war clouds loomed at decade's end.

This *c.* 1935 aerial view of the Portsmouth Navy Yard shows the following, from left to right: Dennetts Island, acquired in 1800, with its industrial shops, building ways, officers' quarters, and (below, to the left of the island's triangular projection) the shorter Dry Dock No. 1; and Seaveys Island, purchased in the 1860s, with its additional shops, tank farm, and (upper right) naval prison and annexes. Jenkins Gut, once separating the two islands, was developed into the longer Dry Dock No. 2 in the early 20th century. (NA.)

The first of three warships, *Falkland*, 54 guns, was built in Portsmouth in 1690 for the Royal Navy. After the reduction of the fourth-rate frigate to 48 guns, she saw use well into the 18th century. In this rendering by George Campbell, *Falkland* comes about on the lee tack. In 1696, the fifth-rater *Bedford Galley*, 32 guns, and more than a half-century thereafter, the frigate *America*, 44 guns, were launched from the Piscataqua River yard. (NH.)

Capt. John Paul Jones received the honor of first raising the Stars and Stripes on an American warship, the Portsmouth-built sloop-of-war *Ranger*, 18 guns. On April 24, 1778, *Ranger* engaged the British sloop-of-war *Drake*, 20 guns, off the coast of Ireland. After suffering heavy casualties, including the mortal wounding of its captain and first lieutenant, *Drake* struck its colors. The casualties aboard *Ranger* were two killed and four wounded. (NH.)

The feisty John Paul Jones, a gardener's son, was born John Paul in 1747, at Kirkbean on Solway Firth, Scotland. Apprenticed to the sea at age 12, he became captain of a brig nine years later. Jones won his greatest fame on September 23, 1779, while in command of the American warship *Bon Homme Richard* during her victorious fight with HMS *Serapis*, 44 guns, when he reportedly proclaimed, "I have not yet begun to fight!" (NH.)

Capt. (later Commo.) Isaac Hull (1773–1843) became the first naval commandant of the Portsmouth Navy Yard in 1813. The previous year, Hull had commanded the frigate *Constitution* in her victory over HMS *Guerriere*—a triumph that earned *Constitution* her enduring nickname, "Old Ironsides." The commandant's wartime tenure saw the launching of the 74-gun ship-of-the-line USS *Washington*. Gilbert Stuart rendered this original portrait, which has since been redone by several other painters. (PS.)

John Locke, a local joiner, built Quarters A, or the commandant's house, during the tenure of the Portsmouth Navy Yard's first commandant, Capt. Isaac Hull, who headed the facility from 1813 to 1815. The oldest extant building in the shipyard, the house is one of the many yard structures that has been elevated to the National Register of Historic Places. (NH.)

This painting, depicting Portsmouth Harbor east of the navy yard, was done by an unknown artist in the mid-19th century. One shiphouse faces the main channel of the Piscataqua River and the other the back channel, dating this quaint rendering from at least 1838. Small crews in dories and shallops busily take fish and lobsters from the inner harbor. (PS.)

Sail sloop-of-war USS *Portsmouth* was launched at the Portsmouth Navy Yard in 1843 and served on the West Coast during the Mexican War. In December 1861, she joined the blockade against the Confederacy in the Gulf of Mexico and engaged Forts Jackson and St. Philip during the federal navy's ascent of the Mississippi River below New Orleans. A survey vessel during the mid-1870s, *Portsmouth* was downgraded to a training ship in 1878. She was sold in 1915. (NH.)

To great national outcry, $1,282,000 was expended in 1852 for construction of a floating dry dock at the Portsmouth Navy Yard, including a marine railway with a 0.83% grade. This dock was used to haul the old frigate USS *Constitution* ashore in 1857 for rebuilding. The floating dry dock served until 1901, and was replaced within a few years by the concrete and granite Dry Dock No. 2. *Constitution* served until 1882; two years later, she became the receiving ship at the yard. (NH.)

During the early 1850s, the renowned frigate *Constellation*, launched in 1797, was broken up at Gosport, Virginia. Budgetary sleight-of-hand allowed a sail sloop-of-war with the same name to be built as a "repair" in order to circumvent congressional prohibition on new warship construction. The 1855 sloop *Constellation*, 22 guns, shown at Portsmouth in the late 19th century, now floats in Baltimore Harbor. (NH.)

The steam sloop USS *Kearsarge* was launched at the Portsmouth Navy Yard in September 1861. Crewed largely by New Hampshire men, she joined the blockade off Gibraltar of CSS *Sumter*, which was commanded by Capt. Raphael Semmes, until that Confederate commerce raider was abandoned in December 1862. Subsequently commanded by Capt. John Winslow, *Kearsarge* sought out Semmes's new cruiser, CSS *Alabama*. On June 19, 1864, the Union warship met *Alabama* off Cherbourg, France, and after an hour-long battle, sank her. (NH.)

Following the Civil War and her recommissioning in 1868, USS *Kearsarge* saw long service in the Pacific Ocean that included carrying a scientific party from Nagasaki to Vladivostok for the transit of Venus. She ultimately returned to the East Coast, decommissioning twice—in January 1878 and December 1886—for overhaul at the Portsmouth Navy Yard, while also seeing service in the North Atlantic, Mediterranean, and Caribbean. While en route to Nicaragua in February 1894, she foundered on Roncador Reef, but without loss of life. (PS.)

Adm. David Glasgow Farragut (1801–1870) presumably uttered the legendary fighting words "Damn the torpedoes!" from the deck of his flagship, the screw sloop USS *Hartford*, in August 1864 while leading his squadron through the Confederate minefield that protected Mobile Bay. During a visit to the commandant of the Portsmouth Navy Yard six years later, Farragut, the first man to hold the rank of full admiral in the U.S. Navy, died in Quarters A at the age of 69. (PS.)

Old salts aboard USS *Hartford* pose during the 1876 national centennial with other crew members: the owl and the pussycat. Cats were favorite mascots aboard wooden ships because they were affectionate and, with a litter box, would not dirty the spotless holystoned decks. Furthermore, shipboard rats were always on the menu for both cats and owls. (NH.)

Originally constructed in 1828, the marine barracks at the Portsmouth Navy Yard was enlarged during both the 19th and 20th centuries. An early configuration consisted of two 3-story Greek Revival end buildings (note, however, the Gothic Revival chimney caps), connected by the long two-story barrack with front-facing lower and upper verandas. The marine personnel housed here guarded the shipyard and later operated the naval prison. Sometimes transient shipboard marines were quartered here, too. (PS.)

By the final decades of the 19th century, a third story had been added atop the barrack's connecting middle portion, and the entirety had been whitewashed, providing a homey appearance. For more than a century and a half, the barrack housed the marine garrison protecting the naval shipyard. Although the marines were withdrawn in 1987 and were replaced by civilian guards, the building still occasionally houses marines on a temporary basis. (SB.)

Quarters B and, beyond, Quarters C/D were completed in 1849 and 1834 for ranking officers of the Portsmouth Navy Yard. The structures exhibit a quaint, pastoral look virtually identical to that of other well-maintained, historic dwellings in towns and villages along the East Coast. Many buildings on the naval reservation—which since November 30, 1945, has been designated the Portsmouth Naval Shipyard—are now on the National Register of Historic Places or have been judged eligible for such status. (NH.)

It is not difficult to identify the youthful marine who has drawn sentry duty at the Portsmouth Navy Yard. While his companions in the marine guard may relax for the camera, the armed sentinel must remain vigilant in watching for intruders. By the time of this c. 1880s photograph, the U.S. Marine Corps non-commissioned rank chevrons had been inverted from those of the U.S. Army. (SB.)

The ironclad USS *Galena* of Civil War fame was scrapped after only 10 years of service. The Norfolk Navy Yard built the second warship so named, a screw sloop-of-war, in 1879. *Galena* saw considerable service in foreign waters and, during the 1880s, served frequently as the flagship of the United States naval forces in the Atlantic. While being towed to the Portsmouth Navy Yard in March 1891, *Galena* and her tug grounded on Martha's Vineyard. Although recovered, she was soon sold for scrap. (NH.)

Marines aboard the screw sloop-of-war USS *Galena* display their spit-and-polish best while the ship is docked at the Portsmouth Navy Yard. *Galena* frequently served as the flagship of the Atlantic squadron and thus was no stranger to Portsmouth. During the 1880s, while protecting American lives and property, she saw much use in the South Atlantic and Caribbean, where these marines were deployed on more than one occasion. (PS.)

This distant view of the Portsmouth Navy Yard includes, on the right, the Franklin Shiphouse and the adjacent mast shear legs. It also reveals four vessels at dockside. From left to right are USS *Constellation*, USS *Dale*, USS *Tallapoosa*, and (bow on) the receiving ship *Constitution*. In 1870, the sidewheel gunboat *Tallapoosa* carried Admiral Farragut here before his death, and in 1882, she brought Secretary of the Navy William E. Chandler for an official visit. (PS.)

Armed with rifles and cutlasses, and under the watchful eyes of commissioned officers, the crew of the sail sloop-of-war USS *Constellation* carries out a repel-boarders drill. The year is 1884, and the vessel, built 30 years before, has become obsolete. The days of the steel navy have already commenced; the enemy tactic of swinging aboard for hand-to-hand combat, which these sailors are practicing to thwart, is already a threat of the past. (PS.)

The frigate *Constitution*—the venerable "Old Ironsides" launched in 1797 and victorious in separate actions in 1812 over the British frigates *Guerriere* and *Java*—was brought to the Portsmouth Navy Yard in 1882 and became a receiving ship two years later. This conversion included the replacement of the vessel's three heavy masts with lighter ones. The gallant ship performed this ignominious duty until her 100th anniversary approached, whereupon she was towed from Portsmouth to Boston and later was used as a museum. (KM.)

To rescue U.S. Army Lt. Adolphus Greely's long-stranded Arctic exploration party, New Hampshire–born Secretary of the Navy William Chandler obtained funding to mount an expedition. Two sealing vessels, *Thetis* and *Bear*, both designed to navigate ice-bound waters, were purchased for the difficult undertaking. In June 1884, Greely (at right) and the other six survivors of the original 25-man party were taken off Ellesmere Island, but one died en route after surgery. The others recovered at the Portsmouth Navy Yard. (NH.)

The sail sloop USS *Jamestown* was launched at the Gosport Navy Yard, Virginia, in 1844 and saw duty all over the world, from Ireland to Alaska. During the Civil War, she captured five blockade runners off the East Coast before sailing to the Pacific to patrol against Confederate commerce raiders. Here, she is tied up at the Portsmouth Navy Yard in 1889; the receiving vessel *Constitution* lies astern of her. *Jamestown* burned at the Norfolk Navy Yard in 1913. (NH.)

A cutter puts ashore at Fourtree Island in Portsmouth Harbor during the late 19th century. The man seated in the stern may be either a river pilot or a naval officer in mufti, perhaps the captain of a man-of-war. Visible up the Piscataqua River are two of the three massive shiphouses and other sizable buildings at the Portsmouth Navy Yard. (SB.)

The double-turreted monitor USS *Agamenticus* was built at the Portsmouth Navy Yard in 1862. After seeing brief wartime service in New England waters, she was decommissioned at the Boston Navy Yard in 1865 and renamed *Terror* four years later. One of the monitors entirely rebuilt under the fiction of "repairs" between 1874 and 1883, USS *Terror* subsequently received a new hull and engines, and two modern 10-inch breechloading rifles were mounted in each of her two turrets. (PS.)

Among the last of the navy's wooden cruisers, the wartime-authorized screw steamer *Essex* was not actually built until 1874, in East Boston's Kitter Shipyard. She served mostly in Asiatic waters during the 1880s, and was thereafter a training ship. *Essex* decommissioned at the Portsmouth Navy Yard in 1898, and is shown at that time near Shiphouses 4 and 5. She then served on the Great Lakes with Ohio and Minnesota naval reserve units, before being burned in 1931. (NH.)

Sailors were once referred to as "iron men in wooden ships," but these 1880s yard workers also merit that description. Posed aboard USS *Marion* with the identifying tools of their skilled trades, these craftsmen toiled outside in all weather, and in unheated shiphouses, 10 hours a day, sawing, smithing, bending, lifting, and fastening the heavy timbers and iron fittings of the warships. (SB.)

Experience, pride, and hard work show in the age, demeanor, and leather aprons of these 1880s foundry workers. The coal at their feet indicates that they are outside of the melting furnace area, a hot and dangerous section where burns from splattering iron were common. Their foreman and sub-foreman are easy to identify from their clothes; judging from his pose, the latter is perhaps contemplating a quick promotion once his gray-bearded boss retires. (PS.)

Three

Concrete and Steel

The Early Modern Defenses

The years immediately following the American Civil War were quiet ones for the military. Not only was the threat distant, but the technology of warfare was rapidly changing as well. Congress wisely decided to forgo new investments in fortifications until new rifled steel guns were adequately developed and plans for new types of defenses had been worked out. By the late 1880s, sufficient progress had been made on the development front. The country was also feeling a new sense of international competitiveness that resulted in the Spanish-American War and the acquisition of new territories. Three important periods of investment in new fortifications followed: the Endicott era (roughly 1888–1906), the Taft era (1906–1915), and the Board of Review era (1915–1925). Portsmouth's defenses benefited significantly during this age of concrete and steel.

New types of batteries, utilizing both guns and submarine mines, were placed at three fortification sites. In addition to retaining the Fort Constitution reservation, new forts were developed at Jerrys Point in New Castle, New Hampshire (Fort Stark), and on Gerrish Island in Kittery Point, Maine (Fort Foster). Later, a post specifically designed as a garrison reservation for coast artillery was also located on New Castle Island. For most of this time, the coast artillery service was a well-trained and adequately deployed force. Over time, thousands of members came and went in Portsmouth, and became a vital factor in the social structure of the city. As this was also the age of affordable photography, we are left today with comprehensive visual records in black and white, both official and amateur.

Ever ready to defend the city and its militarily critical Portsmouth Navy Yard, the army twice rose to the level of active defense during this period. In both the Spanish-American War and World War I, the guns and mines were readied, but no enemy attempted to test American coast defenses. After World War I, these defenses once again became somnolent until called upon for one more emergency.

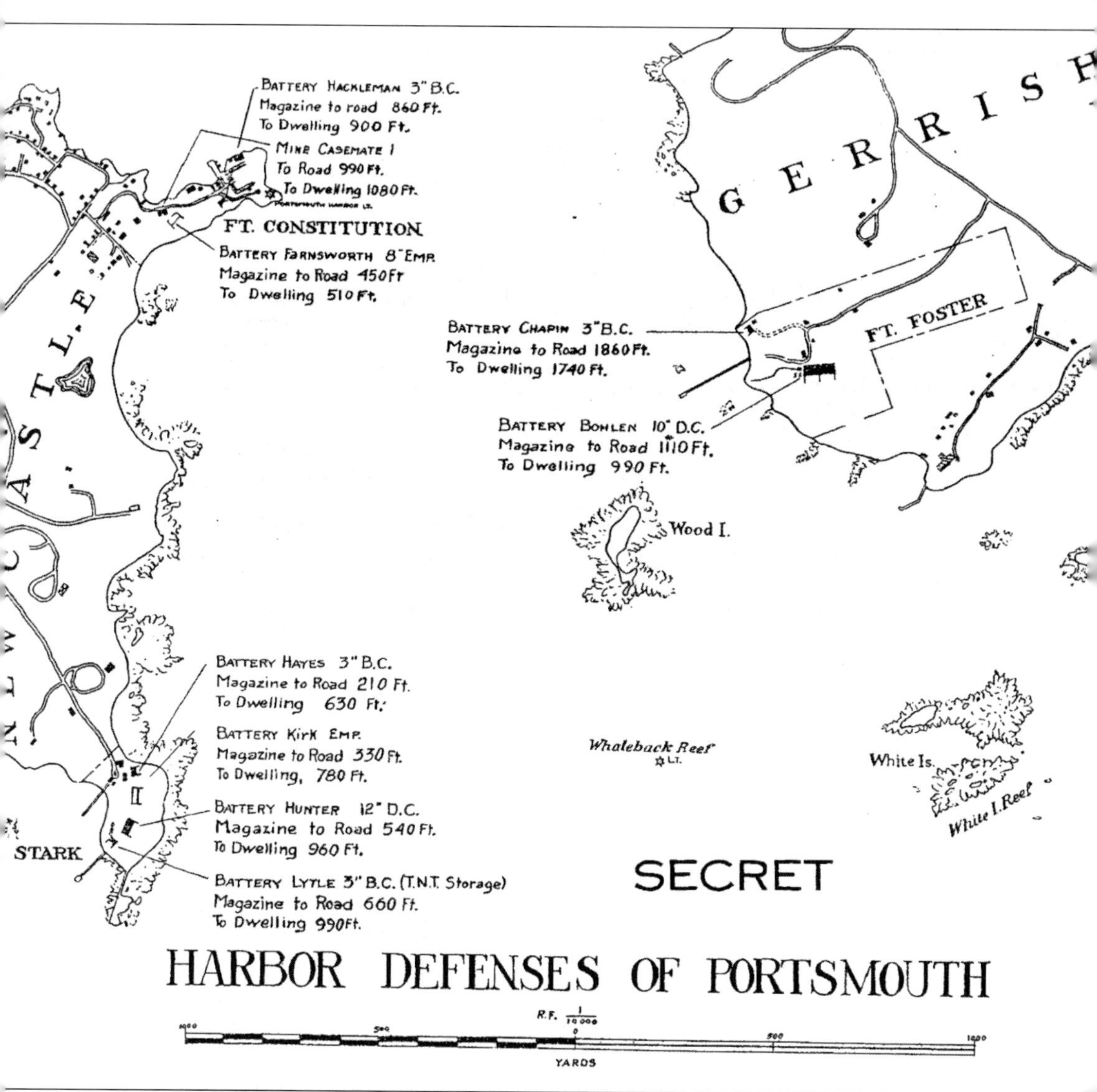

This army drawing of the fortifications of Portsmouth Harbor reveals the excellent position of the three forts developed for Endicott-era batteries. Fort Constitution (upper left), Fort Stark (lower left), and Fort Foster (upper right) are deployed for crossfire in the lethal "triangle of death" around the harbor's main channel. Each fort contained at least one heavy gun battery for engaging major enemy warships, and lighter guns to protect the minefields and defend against smaller vessels. (NA.)

The first significant new gun emplacement for the defense of Portsmouth Harbor was Battery Farnsworth, to the rear of old Fort Constitution. The construction was planned in 1897 and was begun later that year. It was armed with two modern 8-inch guns on disappearing carriages. The battery served until it was disarmed in 1917, during World War I. This recent aerial photograph shows the two gun emplacements in the center left, where the ordnance was once mounted. (GW.)

Local engineer officers designed the fortifications built during the Endicott era based on general specifications provided by the Corps of Engineers. Lt. Col. Andrew M. Damrell (1840–1909), shown here, designed Battery Farnsworth and its adjacent submarine mining casemate in 1897. He was a career officer, graduating from West Point in 1864 and immediately seeing combat service. After leaving Portsmouth in late 1897, he finished his career in the Mobile, Alabama, lighthouse district. (EM.)

This photograph shows one of the guns of Battery Farnsworth, construction of which had just recently completed between 1898 and 1900. The battery, quickly armed at the start of the Spanish-American War, was the only modern facility available for service during that conflict. The gun is in its raised, firing position, with its muzzle just peeking over the concrete parapet toward a potential enemy. On the lower right, some brickwork of the old 1814 Walbach Tower is visible. (BS.)

The proximity of Colonel Walbach's Tower to Battery Farnsworth is evident in this 1903 photograph. In fact, the continued retention of the tower and, more importantly, the substantial rock ledge it sat upon prompted one inspecting officer to call the position a "death trap." Though at the time the army wished to remove the tower, lack of funds and strenuous opposition by the local chapter of the Colonial Dames of America prevented any action from being taken. (NA.)

Battery Farnsworth was unfortunately built with cheaper rosendale cement on the rocky ledge, from which water constantly percolated into the battery. The magazines and interior rooms were constantly damp. By 1905, the battery was in unacceptable condition, and 12 years later, the ordnance was removed for use elsewhere. Photographs of the battery are scarce and, while this 1903 image is of poor quality, it shows a rare frontal view of one of the guns up close. (NA.)

Fort Constitution's second battery in the Endicott era was for light, rapid-fire armament. Battery Hackleman was built in 1904 between Battery Farnsworth and the south bastion of the old fort. It was armed with two 3-inch guns on pedestal mounts with light splinter shields. In this early 1960s aerial view, the emplacement can be seen in the center. The emplacement was entirely removed a few years later. (NW.)

Even in their incomplete states, the seacoast forts continued with a modicum of routine function. Occasionally, the caretakers reshuffled the few remaining pieces of ordnance. In this c. 1900 photograph, a rifled Parrott gun is attached to a sling beneath a transport rig. In the distance behind, several other barrels are visible. Certainly obsolete 40 years after construction, these types of guns were all that was available until the start of the modern Endicott era. (BS.)

The Congressional Report of the Board on Fortification or Other Defenses in 1886 prompted the nation to enhance its seacoast defenses. With the renewed funding of fortification expenditures of the late 1890s, crews returned to begin constructing again. Here, workers at Fort Constitution appear with wheelbarrows and construction implements c. 1900. (BS.)

The Civil War had marked the end of exposed masonry in fortifications. Powerful new smoothbore guns and particularly effective rifled cannon could make quick work of granite and brick. These blocks of granite once earmarked for the exterior walls of Fort Constitution have become expensive, but substantial, foundations for navigation lights in this c. 1898 view. (BS.)

Except for the short-lived excitement of the Spanish-American War in 1898, activity at the Portsmouth Harbor forts was at a minimum for the remainder of the 19th century. Even though still formally military fortifications, they served almost as community parks. Often just a single ordnance sergeant cared for the guns, gunpowder, equipment, and a kitchen garden on post. Fort McClary was a popular place for a weekend stroll or ride, as these cyclists demonstrate. (BS.)

The spherical objects providing seating for these visitors are 15-inch cannonballs. Originally intended for destructive fire during the Civil War, these 315-pound projectiles could be lofted almost three miles by their Rodman cannon. Even stacking them as illustrated required several strong men and the appropriate gins and slings. This photograph, taken at a Portsmouth fort *c.* 1900, likely shows members of the United States 17th Heavy Artillery. (BS.)

When sitting on cannonballs proved too hard on the rear echelon, more conventional seating on steps—in this instance leading up to the half-century-old blockhouse at Fort McClary—was far more comfortable for these soldiers and their girlfriends. One soldier holds a wigwag flag, used either for cross-harbor communication or for early (and woefully inadequate) fire control for the harbor defense batteries. (NC.)

In what was almost certainly an obligatory photograph during any training season, the members of the 124th Company, U.S. Artillery, array themselves in front of the main gate at Fort Constitution. The number of baseball gloves in evidence is an indication of what was really on their minds on a fine summer day in 1902. (SB.)

The members of the 124th Company, now smartly clad, gather in front of the headquarters building at Fort Constitution. Two commissioned officers appear on the far left of the piazza, and a Model 1861 8-inch siege mortar adorns what passes for a front lawn. It is 1902, and the creation of the coast artillery as a formal, separate establishment remains only five years away. (SB.)

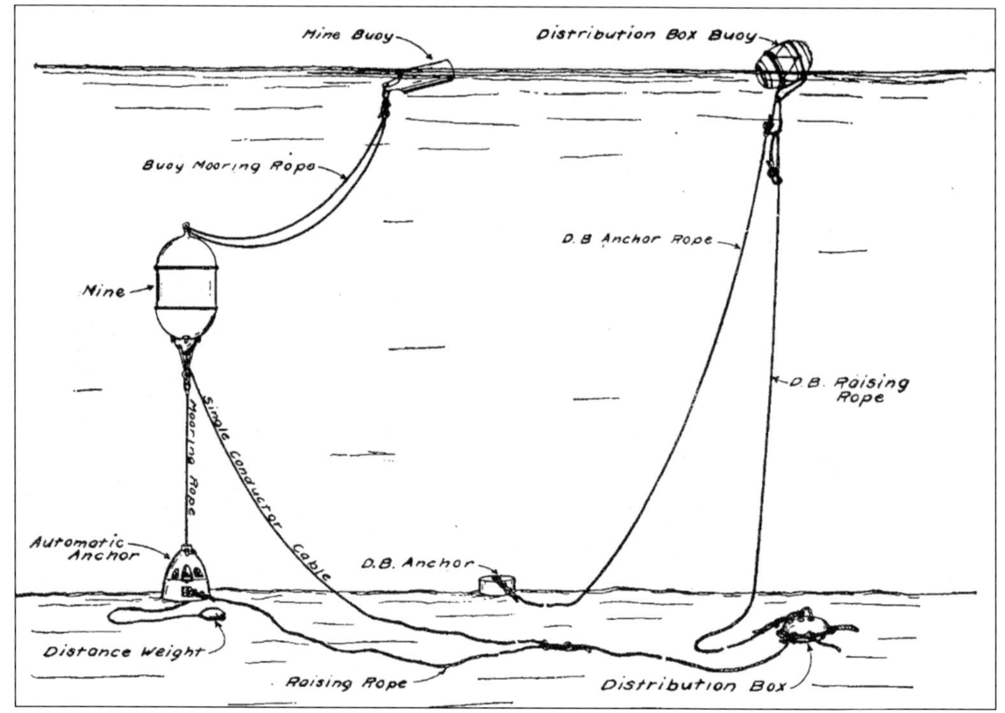

Mines were operated as an integral part of the army's harbor defenses from the late 1890s to the mid-1940s. The most common type was a buoyant mine. It floated at a predetermined depth, was held fast by an iron anchor, and was connected through a distribution box to its electrical cable from ashore. This drawing from a 1912 instruction manual illustrates this typical arrangement. (NA.)

This *c.* 1900 view, taken inside the parade ground, shows some of the submarine mining material at Fort Constitution. The temporary tents denote that a major exercise, or perhaps even troop movement for the Spanish-American War, is under way. The most interesting aspect is the pile of large mine cases. The spherical cases, looking more like buoys than mines, are stored in the open. Within a few years, a storehouse for their protection was erected. (BS.)

Except during practice and actual wartime emergencies, most of the mining material was kept ashore. The cable connecting the mines and the explosive charge were stored in separate structures on post. The spherical mine cases, without explosive, were held in a special torpedo storehouse near the mining wharf. The army continued to use the term "torpedo" for this building despite the widespread use of the term "mine" for the object. This is a 1920s view of the Fort Constitution storehouse. (NA.)

When needed for planting, the mine cases were loaded with explosive. At first, dynamite was used, but TNT was later adopted as the material of choice. Mines were transported ashore on wheeled carts carried on a short run of rails similar to a railway. These rails ran out to the mine wharf, where they came adjacent to the vessels that would carry them to their predetermined planting site. The mine pier at Fort Constitution was built in 1905. (NA.)

On several harbor defense posts, buildings to support the garrison were established as an integral part of the development plans. The army accommodated non-commissioned officers in handsome multifamily quarters. Because professional long-service sergeants were essential to the performance of the artillery, they received preferred treatment. These are such quarters at Fort Constitution. The building was constructed in 1892, and this photograph was taken in the early 1920s. (NA.)

Officers received quarters equivalent to the style and size of housing available to professionals in the civilian world. Unmarried officers were quartered in multi-room bachelor officers' quarters, and often married lieutenants had duplex housing. Senior officers received separate quarters. This family unit for a captain at Fort Constitution was built in the late 1890s. (NA.)

Another necessary post building was the guardhouse. Soldiers were exposed to all of life's temptations, and military rules were usually stricter than civilian ones. The Fort Constitution guardhouse provided facilities for nine guards and four prisoners. Usually, behavior resulting from too much drinking constituted a quick visit or a longer stay in this facility. (NA.)

Unless methodically maintained, wooden buildings do not endure like the stone, brick, and concrete of the batteries. This photograph, a section of a larger panoramic view, shows the lighthouse keeper's residence outside of old Fort Constitution after the end of World War II. The two-story building is starting to show the ravages of time, with peeling paint and missing shingles. Virtually all of the army's wooden buildings on the post were removed as hazards after the war; not a single one remains today. (NC.)

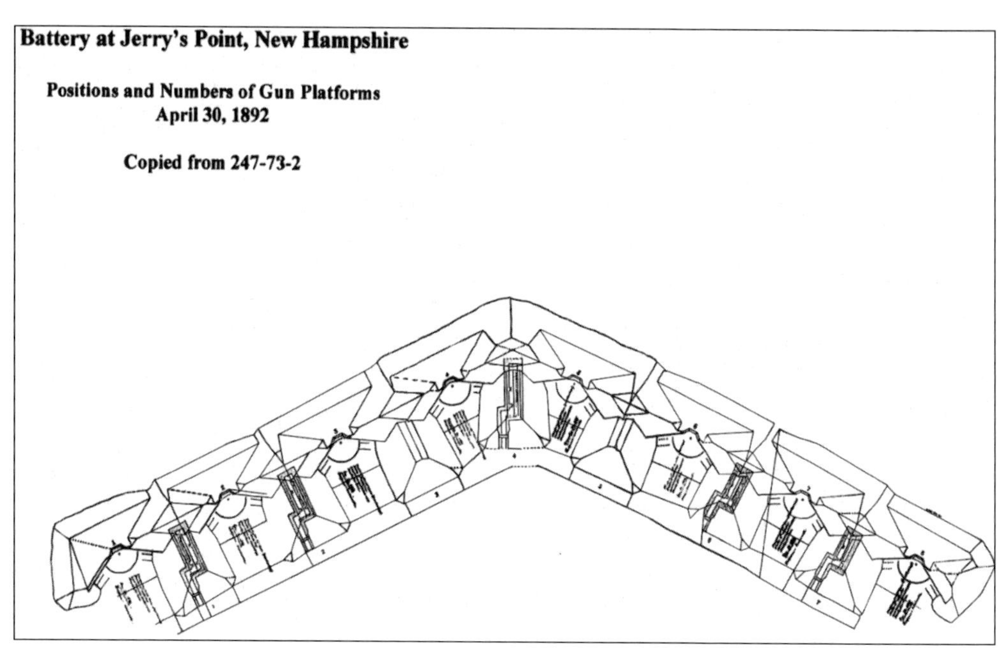

Jerrys Point, on the southwestern tip of New Castle Island, was used for earthen batteries during both the Revolutionary War and the War of 1812. After the federal government acquired the small parcel of land in the 1790s, an arrowhead-shaped earth and concrete battery was begun here in 1873, but it was never completed or armed. This plan from 1892 shows the extent of the work completed before abandonment. The location remained in army hands and was developed into a formidable defensive site in the Endicott era. (NA)

Although this 1880s U.S. lifesaving station was at Jenness Beach in Rye, its counterpart at Jerrys Point served the same function. The station's keeper and assistants used a surf boat to row out to stranded ships in distress, or if close enough, they fired a rope line from the Lyle gun (on the boat ramp) and then rigged a breeches-buoy to carry the sailors ashore. (SB.)

The heaviest battery built for Portsmouth in the Endicott era was emplaced in the center of Fort Stark, essentially requiring the destruction of the older 1870s work. Named Battery Hunter, it mounted powerful 12-inch guns that provided the heaviest shell and one of the longest ranges for more than 40 years. Pictured in the center of this closeup aerial photograph, the battery was flanked on both sides by smaller gun emplacements. (GW.)

This is a *c.* 1915 view from immediately behind one of Battery Hunter's two 12-inch guns. The counterweight of the disappearing carriage has been lowered, bringing the gun up to its firing position. Ammunition was raised from the magazines on the lower level of the emplacement. Usually, mechanical ammunition lifts were used for this movement, but emergency cranes were also available, and are clearly visible to either side of the gun. (NC.)

In this early 1940s view, the 12-inch gun is in its lowered, loading position. In fact, the breechblock has just been opened in preparation for loading. The platform alongside the gun is a sighting platform. One member of the gun crew would directly observe the target and fall of a shot and aim the gun using the optical sight mounted on the platform. (NC.)

The large disappearing guns were supplied with projectiles and powder stored in heavily protected magazines in the lower level of the battery. Projectiles were delivered to the loading platforms by chain hoists, and then transported to the gun breech by carts. These fellows pose with a pair of shell carts carrying 1,000-pound 12-inch shells destined for one of the guns of Battery Hunter at Fort Stark. (NC.)

An Endicott post, Fort Foster, was also established on the north side of the Piscataqua River at Gerrish Island. Utilizing the reservation that had been purchased in 1873, work progressed for 10 years, beginning in 1898, in building two new batteries. The largest was for three 10-inch disappearing guns. The construction of this emplacement, named Battery Bohlen, entailed the destruction of the incomplete 1870s brick-and-concrete battery. This modern aerial photograph shows the battery and its proximity to the river. (GW.)

This is a good illustration of an Endicott disappearing gun, the specific model gun and carriage mounted in Battery Bohlen. The gun is in its raised, firing position. This weapon could fire a 617-pound steel projectile a maximum of eight miles. While adequate when produced in 1901, improvements in battleship gun range soon led to the gun's obsolescence, or at least its relegation to a secondary role. (GW.)

Conditions at Fort Foster were not favorable for a major cantonment. The site was low-lying, marshy ground, and freshwater wells were not available. Only a few camp buildings were built prior to World War II. This is an NCO quarters, built in 1917 for a cost of $2,280. Usually the fort garrison consisted of just a few caretaker sergeants who would have lived in houses on post like this. This particular wooden structure did not survive well; it was razed in February 1944. (NA.)

One of several national guard armories built in the Granite State just prior to World War I is shown in Portsmouth just after construction in 1914. The martial architecture of these buildings (the crenelated battlements and tower) suggests their formidable purpose. In the interwar years, the Portsmouth armory housed one or more units of the 197th Coast Artillery, an antiaircraft gun regiment. In 1940, the regiment was mobilized and eventually was sent to the Pacific theater. (SB.)

Four

Steel Hulls and Steel Guns

The Navy Yard, Part II

During the Spanish-American War—resolved a great deal more quickly than the question of what, or who, was responsible for sinking the battleship USS *Maine*—a prisoner-of-war camp was established on Seaveys Island for Spanish naval personnel, mainly survivors of the battle of Santiago. Arriving in July 1898, the prisoners remained in Camp Long for two months, with their wounded and sick cared for by American Red Cross volunteers.

Commanded by Lt. Col. Robert Huntington, the 1st Marine Battalion guarded the camp. Like the prisoners of war in Camp Long, the marines initially lived under canvas at adjacent Camp Heywood, while waiting for the construction of permanent quarters. In September, the Spanish sailors returned home aboard SS *City of Rome*, the longest liner to have visited Portsmouth Harbor. Because of the number of men still convalescing, five Red Cross nurses volunteered to accompany them as medical orderlies.

Long contemplated, the construction of a large new dry dock in Jenkins Gut, between Dennetts and Seaveys Islands, commenced in 1899. At a little more than $1 million, it cost less than its floating predecessor, built almost a half century before. A few years after the dry dock's completion, the battleship *New Hampshire* was overhauled there.

During 1905, a contractor removed Hendersons Point, a notorious navigational hazard in the approaches to the navy yard, first by building an enormous cofferdam and doing massive excavation, and then by setting off 46 tons of dynamite to blast away the remainder of the granite ledge. Soon thereafter, the emissaries for the peace conference to end the Russo-Japanese War arrived in Portsmouth. At Seaveys Island, construction began on a new naval prison where the Spanish camp had been. When completed, the prison eventually replaced two prison vessels that for a long time had been tied up at the yard to house naval detentioners.

A new, much larger, and more efficient naval hospital was completed in 1913 at a cost of about $3 million; surgical practice in the hospital commenced the next year. The launching of the first yard-built submarine, *L-8*, took place in August 1917, after the United States had entered World War I. Women were hired in substantial numbers during the war, and both civil employees and naval yeowomen were assigned to do exacting shop work. The construction of several O-boats and S-boats began during the war, and submarine building and testing continued through the interwar years.

It is August 13, 1897; unbeknownst to these sailors marching down State Street in Portsmouth, they will soon face death or dreadful harm in battle. But for now, it is a lovely summer's day. The aspect of the parade from the second story of the Central Steam Laundry, while inspiring, must have been a little warm. (SB.)

With war against Spain declared in 1898, the U.S. Navy purchased or leased a number of commercial transports for use as auxiliary cruisers, ideal for scouting. SS *St. Louis* was an 1895 liner leased for such service, although it appears she was used mostly for troop transport. On July 10, 1898, the now-armed USS *St. Louis* steams past the long-abandoned granite walls of Fort Constitution, and the enthusiastic scrutiny of its garrison, to deliver 692 Spanish naval prisoners to Seaveys Island. (BS.)

Another such auxiliary vessel was the 10,600-ton clipper-bow SS *New York* of the American Line, which had set a record time for passage between Southampton, England, and New York five years previous. Armed and redesignated USS *Harvard*, she undertook vigorous lifesaving efforts after the Spanish squadron sortied and met destruction off Santiago, Cuba. *Harvard* saved some 600 Spanish naval personnel and, on July 15, 1898, brought more than 1,000 of these prisoners to the Portsmouth Navy Yard. (NH.)

With spectator boats seemingly ever present, the Spanish sailors were debarked from the auxiliary cruisers–turned–troop transports USS *Harvard* and USS *St. Louis* onto large lighters with names such as *Kittery*, *Durham*, *Eliot*, and *Berwick*. Here, the lighter *Kittery* prepares to cast off from *Harvard* with a load of prisoners, and with the assistance of a tug, will take them to Seaveys Island and the prisoner-of-war camp established there. (NH.)

The Spanish prisoners included those who were wounded in action or ill from disease, some of them gravely. At Portsmouth, 31 died and were buried at the navy yard until their bodies were returned to Spain in 1916. The male nurses present were all American Red Cross personnel who had volunteered to attend to the prisoners. They taught the Spanish medical orderlies, who could better converse with their bedridden shipmates, up-to-date techniques in caring for their patients. (PS.)

Captured Spanish naval officers pose in front of a new building at Camp Long, wearing a variety of garb, some nautical, but most entirely civilian. Even in Portsmouth, New Hampshire, the popularity of Spanish dress style clearly endures. Two Roman Catholic chaplains are in evidence among those seated. In conformance with 19th-century protocol, the officers were paroled under gentlemanly honor and provided pocket money so they could enjoy leisurely jaunts into the city. (SB.)

The 1st Marine Battalion, a provisional unit commanded by Lt. Col. Robert Huntington, had secured the Spanish base at Guantanamo Bay, Cuba, that was needed by the navy. After the fighting had ended, the unit was transferred to Seaveys Island to undertake guard duty. This view shows the battalion formation, with the tents of Camp Heywood behind. Until prisoners were more permanently housed, international treaty obliged the marines to be similarly sheltered under canvas and to eat food identical to that served the Spanish. (NH.)

During the summer of 1898, a marine platoon poses for the camera at Camp Heywood, named for the marine corps commandant, Col. Charles Heywood. The platoon leader, a lieutenant, wears the shell jacket adopted from the British and carries the marine officers' Mameluke sword. His men wear the standard blue blouses, khaki trousers, and leggings, and carry the Lee straight-pull bolt-action rifle, which fired a .236-caliber round. (SB.)

At Camp Heywood, Lt. Col. Robert Huntington sits astride Old Tom, the steed he had ridden at Guantanamo Bay, Cuba. He did so again during the victory parade of the 1st Marine Battalion in downtown Portsmouth on September 16, 1898, four days after the unit's responsibilities at the camp had ended. Upon Old Tom's death in 1933, he was buried at the Portsmouth Navy Yard. (NH.)

Marine officers of the 1st Battalion amuse two ladies, doubtless with hair-raising tales, outside one of their temporary quarters at Camp Heywood, Seaveys Island. These officers, like their men, were obliged to live in identical quarters as the Spanish prisoners they were guarding until the buildings at Camp Long were completed, ideally before the New Hampshire winter. Note the variety of officers' footwear in evidence here. (NH.)

After two months in Camp Long, the prisoners went home. Even before the treaty was negotiated, the Spanish government obtained their release and chartered SS *City of Rome* to repatriate some 1,700 naval prisoners. The elegant clipper-bow liner—560 feet long and weighing 8,415 tons—was built of iron in 1881. *City of Rome* served as a British troopship during the Boer War before being scrapped in 1902. (SB.)

Gunner S. J. Skau (sometimes identified as Skaw or Shaw) poses by his 5-inch gun, reportedly the one that opened battle at Manila Bay against the Spanish squadron anchored along the Cavite shore of Luzon. Historical accounts, including that by the victorious American commander, Commo. George Dewey, specify that the forward main 8-inch gun battery of his protected cruiser, USS *Olympia*, fired the first shots. *Olympia* is now preserved as a floating memorial in Philadelphia. (NH.)

Armstrong-built in 1886 for the Spanish navy, the small protected cruisers *Isla de Luzon* and *Isla de Cuba*, each armed with four 6-inch and four 6-pounder guns, were sunk during the battle of Manila Bay on May 1, 1898. Thereafter, the U.S. Navy recovered and repaired the vessels, and reclassified both as gunboats. *Isla de Cuba* rendered distinguished service during the Philippine Insurrection of 1900–1901 and, three years later, departed for the United States. Here, she is tied up at the Portsmouth Navy Yard. (PS.)

The gunboat USS *Castine* was out of commission at the Portsmouth Navy Yard from 1905 to 1908. Launched by the Bath Iron Works in 1892, the warship mounted eight 4-inch and four 6-pounder guns. She saw active service during both the Spanish-American War and the Philippine Insurrection, and in World War I she served as the tender for the Atlantic Submarine Flotilla. Behind her lies the receiving ship *Constitution*. (PS.)

The armored cruiser USS *Brooklyn*, commissioned in 1896, became Commo. Winfield Schley's flagship and assumed the lead role in the battle off Santiago, Cuba, in July 1898. She visited Portsmouth in the early fall of that year. As the flagship of the Asiatic squadron in 1900, she participated in the China Relief Expedition and, five years later, carried the remains of John Paul Jones from France to Annapolis. *Brooklyn* was sold in 1921. (PS.)

These posed marines, part of the contingent aboard USS *Brooklyn* in 1898, prepare to sell themselves dearly in the defense of Seaveys Island, at least for the camera. The rifles displayed are .236-caliber Lee straight-pull, bolt-action military rifles, 10,000 of which had been manufactured for the U.S. Navy by the Winchester Repeating Arms Company. Although the idea was sound, the guns' design was imperfect, and they were soon replaced by the army M1903 Springfield. (PS.)

To replace the floating timber dry dock in use at the Portsmouth Navy Yard since 1852, between 1900 and 1904 the John Pierce Company of New York built a 750-foot-long dry dock in Jenkins Gut, between Dennetts and Seaveys Islands. Here, *c.* 1901, laborers and the overhead traveling bridge crane excavate the area for Dry Dock No. 2. The massive construction project cost more than $1 million—$193,000 less than its floating predecessor. (NH.)

The side of the yard's massive dry dock was half completed by June 1902. It was built of stepped granite blocks that followed the hull contour of a battleship. Before the area was dug out, a cofferdam was constructed out into the Piscataqua to keep the excavation dry. This also allowed the dock's gates to be fitted when the stone walls were complete. The sides were stepped so that timbers could be wedged against a ship to keep it upright when the water flowed in or was pumped out. (NA.)

The new dry dock is nearing completion in this *c.* 1903 photograph, which shows one of the yard's large railway cranes. The project required the removal of 166,000 cubic yards of ledge, followed by the pouring of 18,000 cubic yards of portland cement concrete for the base and 20,500 cubic yards of ashlar blocks for the floor and sides of the dry dock. The artillery pieces straddling the flagpole are Mark IV 3-inch landing guns. (NA.)

The just-completed Dry Dock No. 2 will soon be full of vessels under repair, but in this 1904 photograph, it is devoid of ships, workers, and bridge crane. The large railway crane appears on the left, and the long stone building on the right is the electrical shop and sail loft. (PS.)

Three naval vessels of different types are being worked on in the huge Dry Dock No. 2 at Portsmouth Navy Yard *c.* 1908. The view is from the head end, nearest to where the tugboat USS *Uncas* is under repair. The dock was constructed of sufficient length to contain the largest conceivable battleship or several smaller ships at once. As revealed on the right, horse drayage was still largely relied on for delivery tasks then. (KM.)

Here, Dry Dock No. 2 holds three other vessels, although this photograph was taken from the river end of the enclosure. As does the previous shot, this one shows clearly the gangways and stabilizing hull timbers within the dry dock. It also reveals the various machine and sheet metal–working shops on Dennetts Island, built close to the dock for which they provided materials. (KM.)

Because the approach to the yard was hindered by Hendersons Point, a 500-foot rocky spur, a congressional appropriation in 1902 permitted the navy to hire the Massachusetts Contracting Company to demolish that navigational hazard. Erection of a huge cofferdam at the point, extensive blasting, and rock removal by a narrow-gauge railway followed, with the final act an immense fulmination of dynamite. (PS.)

Three weeks before the final detonation, the cofferdam stood far above the bottom of the excavation on Hendersons Point. In this photograph, dated July 1, 1905, the vessel on the Piscataqua River literally towers over the tiny figures of men below. The excavated material was removed in strings of narrow-gauge dump cars pulled by steam locomotives. (PS.)

As a crowd of more than 18,000 spectators watched from safe vantage points, at 4:11 p.m. on July 22, 1905, Miss Edith Foster triggered the current to the detonator caps, and 46 tons of dynamite blasted skyward, demolishing what remained of Hendersons Point. Rear Adm. William W. Mead had ordered the river closed, and earlier had issued a warning to the residents of waterfront towns to take down, as he put it, valuable "bric-a-brac" from shelves and mantles. (PS.)

Shortly following the gigantic blast that destroyed Hendersons Point, river launches moved into the area, inspecting the floating debris that had resulted from the explosion. It was reportedly the largest explosive charge that had yet been detonated anywhere. A standpipe can be seen in the distance on Seaveys Island, underscoring the lack of freshwater at that place. (NH.)

In 1905, Japanese (left) and Russian envoys met at the yard in Building 86, the General Stores Building—thereafter known as the Treaty Building—in order to negotiate the end of the Russo-Japanese War, which had been costly in both treasure and blood. When meetings were not in session, the delegates stayed at the Wentworth by the Sea hotel in New Castle. The peace deliberations were arranged by Pres. Theodore Roosevelt, whose active role earned him a Nobel Peace Prize. (PS.)

The signing of the Treaty of Portsmouth that had ended the Russo-Japanese War one year previous, is celebrated on September 4, 1906, by the laying of a plaque outside of Building 86. Note the formation of marines drawn up and the flags of the three nations involved waving over the crowd. Although automobiles did take the envoys about, horse-drawn transport is the only means visible on this day. (PS.)

In July 1903, the prison vessel *Southery* arrived at the yard, marking the transfer of the naval prison from Boston. The construction of the U.S. Naval Prison at Portsmouth began that year; however, seven years after the prison's completion in 1908, *Southery* and *Topeka* (also used as a prison ship) were still docked at the yard to house naval detentioners, so overcrowded had the main cellhouse become. (PS.)

Detentioners confined on the prison ships *Southery* (left) and *Topeka* fall in on the dock for morning roll call. The sailors on the right appear to have earned the privilege to wear naval uniform again, either because they will be released soon or have already accomplished that end by diligence and good work. The next activity on the dock will be morning calisthenics to keep the prisoners physically fit. (NH.)

The navy established a prison on the southern end of Seaveys Island in 1908 to house sailors who had received court-martial sentences for major infractions. The original long cellhouse is on the left of the old administrative tower block, with later cellhouse additions to its right and left. The tower contained offices, an infirmary, recreation rooms, a chapel, and dormitories for sailors who did not warrant cell confinement. There was no fence, but marine sentries patrolled a "dead line," a marked area around the buildings. (PS.)

The naval prison included 320 single-person cells arranged on four tiers inside the original cellhouse. Each cell measured 5 feet 7 inches by nearly 10 feet and contained a strap-iron bunk folding against the wall, a sink, and a toilet. The extra beds along the wall indicate overcrowding in the prison; temporary wooden buildings were erected to house 2,300 additional prisoners in World War I. (PS.)

Marine guards, bundled against a cold wind in the prison yard, assemble *c.* 1910. These men escorted work details of prisoners to various jobs in the navy yard and kept watch along the dead line. If a prisoner crossed this line, he was warned once by a guard before being fired upon. Though not true, marine guards were told that if they allowed a prisoner to escape without shooting, they would have to serve the remainder of the prisoner's sentence. (NH.)

A detentioner's life at the naval prison was hardly spent sitting around in his cell. Work details at the yard occupied many hours a day. In this *c.* 1914 photograph, seven prisoners, accompanied by their marine guard, prepare to haul their cart from the jailhouse yard. The cart likely contained slops to be dumped into the Piscataqua. (PS.)

The battleship USS *Kearsarge*, named for the famed Civil War sloop, is shown anchored in the lower Piscataqua River during the early 20th century. Launched in 1898 and commissioned two years later, *Kearsarge* served as the flagship of the North Atlantic squadron. In 1906, she suffered a gunpowder accident that killed 10 men; nonetheless, *Kearsarge* departed the following year with the Great White Fleet for its voyage around the world. After a six-year modernization, she was recommissioned in 1915. (SB.)

Kearsarge mounted her four main 13-inch guns and four secondary 8-inch guns in odd, superimposed turret pairs. Shown in Portsmouth during World War I, she had by this time been relegated to a gunnery and engineering training ship. In 1920, she was decommissioned and a 250-ton-capacity revolving crane was mounted on her deck. *Kearsarge* steamed to Portsmouth in September 1939 to raise the sunken submarine *Squalus*, and then as *Crane Ship No. 1*, she sailored on during the war. (PS.)

As did its urban counterparts, the Portsmouth Navy Yard fire brigade took great pride in the appearance of its apparatus. The diligent firemen and the burnished metalwork on these engines conjure up images of lathered teams galloping hellbent-for-leather to a fire, smoke belching from the clattering pumpers' boilers. The larger openings on 20th-century fire hydrants, called "steamer connections," are vestiges of the large-diameter suction hoses used by the horse-drawn engines. (SB.)

The yard's central powerhouse, on the left, was enlarged in 1901 to accommodate the needs of the new Dry Dock No. 2, which is seen under construction at the right edge of the photograph, just beyond the rigging storage building. Boilers in the plant produced high-pressure steam, which rotated reciprocating-engine-driven generators to light the facility's many buildings and to power the thousands of machines needed for ship construction and repair. Exhaust steam was carried in underground pipes to heat the buildings and to make hot water. (NA.)

The battleship USS *New Hampshire*, armed with four 12-inch guns and twenty 8-inch and 7-inch guns, departs Portsmouth Harbor after a nine-day stay in 1908. *New Hampshire* had been commissioned only five months previous, too late to take part in the voyage of the Great White Fleet around the world. However, she was present in Hampton Roads, with Pres. Theodore Roosevelt aboard, to welcome the fleet home in 1909. (PS.)

New Hampshire sits in Dry Dock No. 2 during one of two refits at Portsmouth in 1909. As with most period battleships, her pole masts were replaced by cage masts sometime before the United States became involved in World War I. During the war, she served in training and escort roles, undertaking postwar missions to the Caribbean and the Baltic. In 1921, she was decommissioned and thereafter scrapped in compliance with the Washington naval limitations treaty. (PS.)

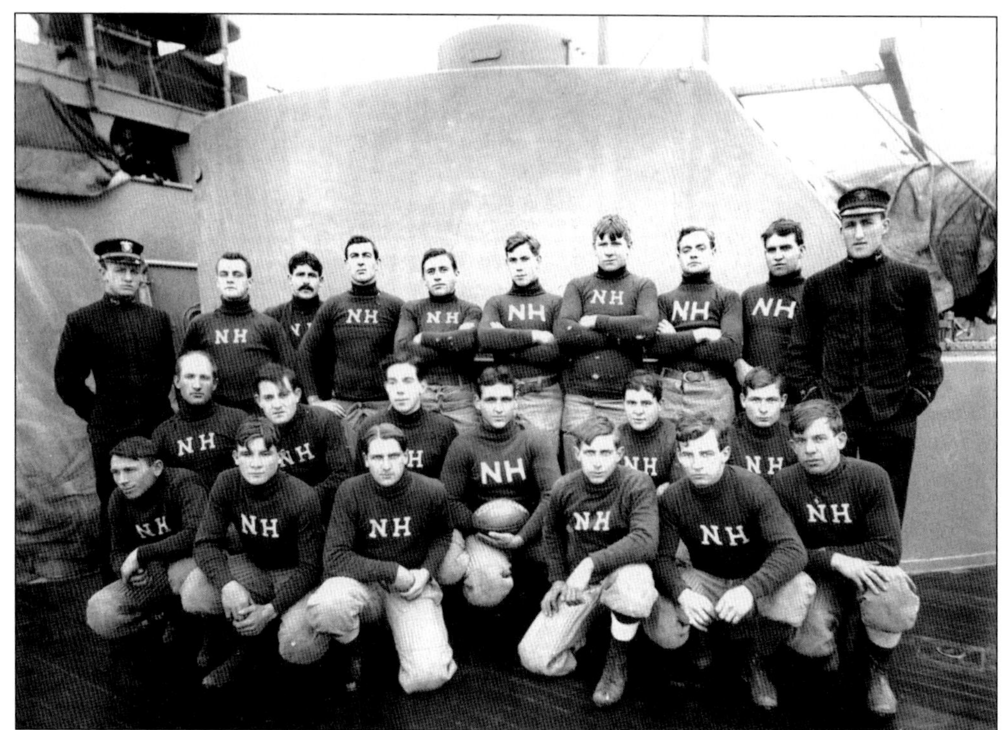

Flanked by their junior officer–coaches, members of *New Hampshire*'s football team display manly demeanor during the battleship's repair in Portsmouth between early October and mid-December 1909. Winning teams brought renown to a ship, extra cash to those swabbies who wagered wisely, and a few black eyes and a visit from the Shore Patrol to those who questioned another ship's qualities. *New Hampshire* visited the Portsmouth Navy Yard four times between August 1908 and July 1910. (SB.)

The armored cruiser USS *North Carolina*, mounting four 10-inch guns in two turrets, steams up the Piscataqua River in August 1911, after bringing home bodies from the USS *Maine* for final interment at Arlington National Cemetery. In 1915, she became the first ship to catapult-launch an aircraft while under way, and during the war, she escorted troop transports. Astern of her, at the head of the long wharf in New Castle, is the still-standing Piscataqua Café. (NH.)

With the newly built naval prison frowning from the headland on Seaveys Island, the equally brand-new scout cruiser USS *Chester*, with two 5-inch guns and six 3-inch guns, leaves Portsmouth Harbor in 1908. In subsequent years, she saw much service in the Caribbean Sea and the Gulf of Mexico and, during World War I, undertook escort duties in both American and European waters. *Chester* was decommissioned in 1921 and sold for scrap nine years later. (PS.)

Armed with tools and pistols, *Chester*'s Pioneer Squad falls in at dockside while their ship visits the Portsmouth Navy Yard in 1913. Such small units were sent ashore to prepare defensive positions or accomplish other labor; their pistols provided self-protection when in unfriendly territory. (PS.)

The handsome hospital at the Portsmouth Navy Yard was built in 1891 and served that purpose until it was replaced by a new medical facility in 1913. Thereafter, the building became the post's bachelor officers' quarters. It was later used for classroom instruction, including the yard's apprenticeship program in the specialized mechanical skills needed to construct steel warships and machinery. (PS.)

After two decades of use and enormous changes in medical science, the old naval hospital no longer met the spatial needs nor the standards demanded by modern medicine. Between 1910 and 1913, a new $3 million hospital was constructed at the yard. By 1914, the hospital was staffed and open, and the first major surgeries were being performed. This 1965 photograph shows the newer annex buildings added during World War II. (NA.)

Handsome brick buildings were constructed to house the yard's many fabricating shops, and the importance of this naval facility was expressed in the architectural style of even its mundane factory structures. A 1915 addition on the left of the foundry building reproduced the window, pilaster, and clerestory details of the original 1860s structure, resulting in a building that both met the yard's engineering needs and continued its heritage. Buildings constructed during the World War II emergency were functional, but hardly architecturally notable. (NA.)

The foundry occupied two stories to accommodate the overhead traveling cranes that moved heavy castings from the furnace and molten-pouring section to cooling tanks, then to the storage area in the foreground. From there, they were transported by rail to a machine shop to be forged, milled, turned, and drilled into thousands of parts for warships. Many different metal alloys were used for these parts, which included the intricately cut, hardened-steel reduction gears and the huge brass propellers, finely balanced to reduce vibration. (NA.)

Shortly after the United States declared war on Germany in April 1917, part of Seaveys Island was set aside as a camp to train naval recruits. These seamen apprentices are members of the 2nd Company, 1st Battalion, and appear to be standing up well to the rigors of boot camp. (NH.)

Needing only George M. Cohan to finesse this production, naval yeowomen pose for a patriotic group photograph in May 1919, six months after the armistice was declared. Some 1,800 women were serving in the U.S. Navy by January of the previous year. (NH.)

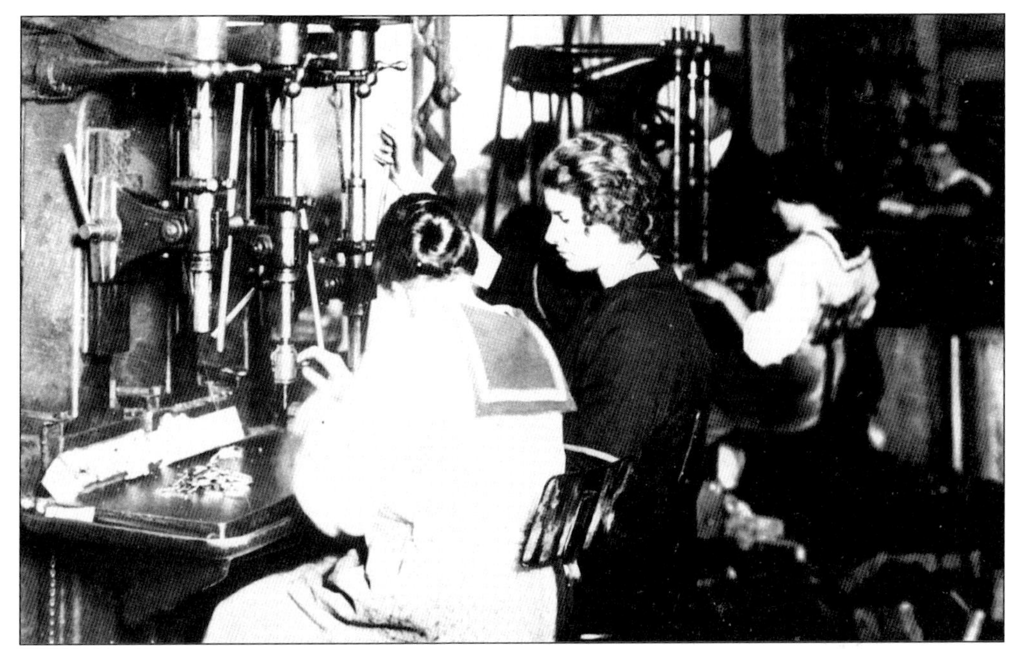

During World War I, naval yeowomen worked essential jobs at the Portsmouth Navy Yard, in many instances performing the same tasks that the female civil employees did. Here, three yeowomen operate drill presses. In less serious but equally rigorous pursuits, they fielded their own rowing team for the 1918 Boston Regatta. (NH.)

Beginning in World War I, Portsmouth-area women were invaluable to the yard's wartime activities. A thousand of them replaced men needed in the military services, and the women provided a source of new workers as shipbuilding greatly increased. Here, women operate stamping machines in one of the machine shops, producing sheet-metal parts formed by dies pushed together by the great pressure of a descending ram. This was dangerous and tiring work, but these ladies did not flinch. Also, the women were absent less frequently than their male counterparts. (PS.)

In a landmark event, on April 23, 1917, the first submarine built by the Portsmouth Navy Yard, *L-8*, was launched from the Franklin Shiphouse. The submarine required more than two years to complete after her keel laying, but only four months between launching and commissioning. Among those on the opposite shore watching the *L-8* slide into the back channel are a handful of soldiers or marines standing on a shed roof. (KM.)

The submarine *L-8* appears in dry dock sometime after launching. She eventually joined Submarine Division 6 in the Azores, but the armistice had been declared by the time she reached Bermuda. Although *L-8* became involved for three years on the Pacific Coast in experimental work with torpedoes and underwater sound detection, it was clear she was primitive when compared with the wartime German boats. She was decommissioned in November 1922 and sold three years later. (NA.)

After the construction of the yard's first submarine, *L-8*, in the Franklin Shiphouse, the engineering staff realized that the shiphouse should be fully converted to a submarine construction facility. In this March 3, 1921 photograph, the shiphouse ways are being altered to accommodate the smaller vessels soon to be built here. (PS.)

The L. H. Shattuck Shipyard was established in Newington, New Hampshire, during World War I in order to build 15 steam-powered wooden merchantmen, each of 3,500 tons. They were clearly stopgap vessels, conceived during the period of urgent need for oceangoing transport. Shattuck launched six ships before the November 1918 armistice. Attended by many spectators, *Roy H. Beattie* slides down the ways, to be followed shortly by *Chibabos* and *Milton*, on July 4, 1918. (SB.)

The 15 contracted 3,500-ton freighters were completed by Shattuck and subsequently were approved by the U.S. Shipping Board. The first one, *Beattie*, launched on July 4, 1918, and the last, *Newburyport*, on August 14, 1919. SS *Newton* is shown just prior to her launch on January 4, 1919. Note the hawsehole for the anchor chain and the plumb marks on the bow. (SB.)

Downriver at Freemans Point (now Atlantic Heights), the Atlantic Corporation built 10 steel merchantmen of 8,800 tons deadweight for the U.S. Shipping Board, a government organization that built 1,000 merchant ships and operated captured German vessels during World War I. In this view, a crowd gathers to watch the second of these vessels, SS *Babboosic*, being launched on May 3, 1919. Its overlapping steel plates and the empty hawseholes, from which the anchors will soon be suspended, are visible. (SB.)

The threat of rain on Memorial Day in 1917 has clearly failed to deter Portsmouth residents from coming out to watch the marines and sailors march down Congress Street. It may, however, explain the anomaly of the marines wearing campaign hats with their blues. Originally called Decoration Day because the graves of Civil War dead were adorned with flowers, the occasion of honoring those who had died while defending this country became a widespread observance after 1868. May 30th was most likely chosen because flowers would be in bloom then. (PS.)

On Armistice Day, ending World War I, a parade started at the yard and crossed the bridge into Kittery, with a navy band in the lead. Many of the shipyard shop and warehouse buildings are visible across Seaveys Island. The young boy in knickerbockers running across in front of the band is doing what all small boys enjoy: something impulsive and unexpected. (PS.)

A pair of World War I–era escort vessels are tied up in the back channel near the Franklin Shiphouse in May 1922. By this date, the shiphouse had become a large, familiar landmark, and its conversion to a submarine-building facility was well under way. Such shiphouses were necessary in cold and snowy climates to permit year-round shipbuilding. (NH.)

Ford Motor Company patrol boats lie mothballed at Portsmouth during the 1920s. In 1917, the navy realized its immediate need for escort vessels smaller than destroyers but having a longer range than the wooden 110-foot submarine chasers. Henry Ford was consulted because of his knowledge of mass-production techniques. He urged that the vessels be built with flat plates, not rolled ones, to ensure rapid production. The 60 resulting Eagle boats had an ugly design and poor seakeeping qualities. (NH)

Five

Submarines and a Two-Ocean War

The Navy Yard, Part III

During the years between the two world wars, the Portsmouth Navy Yard was not inactive and the period was hardly uneventful. In this two-decade span, the yard constructed 20 submarines and one surface vessel, the large coast guard tug *Hudson*—the yard's first such ship in many years and its last until this very day. The vessel was also a first for the Portsmouth Navy Yard in another respect: it was constructed in prefabricated sections, carried between the erecting and assembling sites by the yard's railroad system. The building program accelerated in 1933 as a result of the deteriorating world situation, and three years later, the yard introduced all-welded construction, which was far superior to the former riveted method.

Several vessels of note came calling during the 1920s: USS *Langley*, the navy's first aircraft carrier; and just prior to American involvement in World War II, the monster French submarine-cruiser *Surcouf*, with its pair of 203-millimeter guns. After a major overhaul at the yard, *Surcouf* sailed to Canada, and in joining a small Free French naval expedition to liberate two Vichy French islands off the south coast of Newfoundland, it embarrassed both Canada and the United States. Shortly thereafter, *Surcouf* disappeared under uncertain circumstances in the Caribbean.

These years also brought challenges and catastrophes. Two of these disastrous episodes, the grounding of *S-48* in 1925 and the sinking of *Squalus* in 1939, allowed the navy to exercise its skill and ingenuity in salvaging and repairing these submarines. The third event, the loss of *O-9* with all hands on the eve of war in 1941, was an unmitigated tragedy.

The yard delivered 67 submarines of its own construction during the course of World War II, plus the completion of two boats that had originated in another yard. Most of these large fleet boats went to the Pacific Ocean, where, with a handful of S-boats and V-boats still operational, they wreaked havoc on the Japanese navy and more particularly, on its merchant marine. As the war wound down in 1944, so did submarine construction at the yard. On November 30, 1945, shortly after war's end, the facility was redesignated the Portsmouth Naval Shipyard.

In successive years, the shipyard carried out the conversion of existing fleet submarines to GUPPY—Greater Underwater Propulsive Power—boats, and thereafter the construction of an entirely new type of submersible, powered by nuclear energy. In 2004, the naval facility no longer constructs vessels, but does rebuild, refit, and overhaul the large fleet submarines now integral to the modern U.S. Navy.

More advanced than the L-boats and O-boats, *S-5* launched at Portsmouth one year after World War I had ended. Here, she is seen undergoing shakedown. On September 1, 1920, during a trial dive off the Delaware Capes, water pouring through the main induction valve flooded the boat and caused her to sink to the bottom at nearly 200 feet. With great courage and ingenuity, the crew brought her up, the stern bobbing above the surface. The entire crew was rescued, but *S-5* was not recovered. (NH.)

With the famous Wentworth by the Sea looming beyond the New Castle causeway, two tugs assist the submarine *S-12* just after her launching on August 4, 1921. Much of her subsequent service was spent in the Caribbean Sea, predominantly at the Canal Zone and Guantanamo Bay. Decommissioned in the 1930s, she was recommissioned on the eve of war, and went to the Caribbean. *S-12* was sold for scrap in October 1945. (NH.)

In 1920, the navy collier USS *Jupiter* was converted into the first American aircraft carrier, USS *Langley*. Important trials were carried out using this vessel, including landings and both unassisted and catapult takeoffs. This photograph was taken at Portsmouth during one of her exhibition voyages along the East Coast about 1923 (as evidenced by the new Memorial Bridge in the background). Later relegated to an airplane ferry, *Langley* was lost to Japanese aircraft south of Java in February 1942. (PS.)

Submarines *S-14* and *S-17* undergo tests, including submergence, in the winter of 1921. Both boats were launched shortly after the end of World War I, and both saw extensive service in Asiatic waters and at the Panama Canal. They served in World War II at the Canal Zone and in Casco Bay, Maine. *S-14* was sold for scrap in November 1945, while *S-17* had been intentionally sunk the previous April. (PS.)

S-16 is submergence-tested in the spring of 1921. Tied up opposite her is the German submarine *U-111*, which had arrived at Portsmouth two years before. During the 1920s, *S-16* operated in the Philippines and at the Panama Canal. During World War II, she saw service in the Caribbean, around the Canal Zone, and in Casco Bay. She was intentionally sunk off Key West in April 1945. (PS.)

The venerable Alabama and Santee shiphouses were razed in 1899 and replaced with covered submarine-building ways in 1917. In this 1927 photograph, the very large submarines *V-4* (later named *Argonaut*) and *V-5* (later named *Narwhal*) are under construction within. Although armed with two 6-inch guns each, the submarines were only near-sister boats because *V-4* was built originally as a minelayer, with four torpedo tubes forward and two mine-laying tubes aft. (NH.)

In 1924, the submarine *V-1* (named *Barracuda* seven years later) appears on the building ways at the Portsmouth Navy Yard in both the upper (stern) and lower (bow) views. Although the first three V-boats were beautifully streamlined and sleek in design, in actual use they were underpowered, slow, and unwieldy. Recalled to duty at Portsmouth in 1940, the three subs were sent to relatively safe patrol areas southwest of Panama. In 1942, they returned to New London, Connecticut, and spent the remainder of the war there, until they were sold in 1945. (PS.)

Barracuda's sister, *V-3*, later named *Bonita*, requires the assistance of three tugs to negotiate the narrow channel after her launching at the yard on June 9, 1925. She, too, was assigned to a safe wartime patrol area near the Canal Zone, before being returned to New London. Note the interesting Boulter Company harborside operation in the background. (PS.)

At nearly the same angle, *V-3* is photographed approaching her berth under her own power, now armed with two 6-inch deck guns, a year or so after launching. Whatever they lacked in power and maneuverability, and however much they disappointed expectations, the first three V-boats were beauties. The second three V-boats were larger and more powerful, but no less ungainly and disappointing. (NH.)

With the new Memorial Bridge in the background, the large *V-4*, later named *Argonaut*, ties up just after her November 1927 launching. Her war career proved to be unlucky. Along with near-sister *Nautilus* (*V-6*), she converted to a troop-carrying submarine and participated in the controversial marine commando raid on Makin Island in August 1942. In January 1943, during her first combat patrol in the Solomon Islands, *Argonaut* was sunk with all hands by Japanese destroyers. (PS.)

Sharing Dry Dock No. 2 in March 1928, World War I submarine *O-2* is dwarfed by the younger *V-4*. The larger submarine was only days from her commissioning in early April, and the smaller Puget Sound–built boat was being modernized to return to first-line submarine status. From Portsmouth, *V-4* was assigned to Newport, Rhode Island, and *O-2* to New London, Connecticut, just down Long Island Sound. (PS.)

While nearing the Piscataqua River on the night of January 29, 1925, during a fierce storm, *S-48* grounded first on Jerrys Point, and again inside Little Harbor. The following morning, coast guardsmen rescued the crew, and in early February, a salvage crew freed the submarine for towing into the yard. To haul her into the Franklin Shiphouse two years later, a plan was devised to use three 0-6-0 shifting engines leased from the Boston & Maine Railroad, connected to blocks and a hauling bridle. (NH.)

On February 3, 1927, with triple-headed locomotives providing the motive power on the other end of the cable, *S-48* was winched into the Franklin Shiphouse for extensive repairs. The yard engines were too small to accomplish the task, but with the larger Boston & Maine locomotives, the well-executed maneuver required but 20 minutes. (PS.)

This photograph, taken from the interior of the Franklin Shiphouse, shows *S-48* being hauled onto the ways on February 3, 1927. During the course of the overhaul, the decision was made to lengthen the original 240-foot boat to 265.5 feet. The renovated *S-48* then launched from the shiphouse on September 4, 1928. (PS.)

After recommissioning in December 1928, USS *S-48* became operational, but during test exercises off New London in April 1929, her main electric motor suffered a defect. She returned to dry dock at the Portsmouth Navy Yard, where this photograph was taken. Thereafter, *S-48* operated out of the Canal Zone for several years. During World War II, she alternated with *S-20* as the station boat for antisubmarine warfare exercises in Casco Bay, Maine. In January 1946, *S-48* was sold for scrap. (NH.)

Sisters *O-9* and *O-10* are tied up at Philadelphia after their 1931 decommissioning. Built during World War I at Fore River in Quincy, Massachusetts, they were part of an American submarine flotilla that arrived in European waters just after war's end. Recommissioned in 1941, they were ordered to the Isles of Shoals to conduct deep submergence tests. On June 20, *O-9* failed to surface after her first dive, crushed at a depth exceeding 400 feet. (NH.)

A postwar Omaha-class light cruiser, mounting twelve 6-inch guns in both turrets and casemates, USS *Concord* is tied up at the Portsmouth Navy Yard in the summer of 1934. She was assigned as a convoy escort during 1942, and a year later transported Adm. Richard Byrd in his survey of the southeast Pacific islands. In March 1944, she bombarded the Kurile Islands and disrupted Japan's northern shipping lanes. *Concord* was sold for scrap in January 1947. (PS.)

An 0-4-0 tank engine pushes mobile rail crane No. 5 into position to help lift the stern section of coast guard tug *Hudson* onto a flat car. The lack of a coal tender identifies this locomotive as a "fireless cooker," whose use was common in industrial works, where an open firebox constituted a potential hazard. In 1934, *Hudson* became the first vessel assembled from prefabricated sections and the last surface vessel built by the Portsmouth Navy Yard. (PS.)

The rail supervisor and crew of another 0-4-0 tank engine pose in the midst of their labors. The photograph clearly shows the flathead valves and wooden cab of this old engine, still operating during the 1930s. In the background, a snazzy coupe stands in front of Building 7, the riggers' shop. (PS.)

The massive reduction gears were cut on huge vertical milling machines in another shop, but in this machine shop, hundreds of smaller gears were produced. Pinion gears from these lathes, for example, were used to rotate the gun turrets and bevel gears to open and close steam valves located in difficult-to-reach places. The lathes were belt-powered by a system of overhead shafts and pulleys, turned by a large electric motor. (PS.)

Hot trisodium phosphate and sulphuric acid baths were used to clean metal parts before undergoing electroplating or anodizing to prevent corrosion. Although one of these men is wearing rubber gloves, they are little protected from fumes as they lower a section of piping into the TSP tank in the 1940s. Shipyard employees and sailors who worked with asbestos in the boiler and engine rooms of warships fared worse, and many later died from asbestosis. (PS.)

By 1931, when this view of the blacksmith shop was photographed, at least one woman had taken up her hammer in this traditionally male occupation. Though there are mechanical drop hammers and power shears along the windows, the dirt-floored center aisle of the shop, with its anvils and glowing coal-fired forges, is the center of activity. Many parts still had to be hammered out by hand after initial shaping with the powered equipment, and only a skilled worker could heat, shape, and quench the parts in oil to harden them properly. (PS.)

The 98-year-old Franklin Shiphouse burned to the ground in the early morning of March 10, 1936. Despite strenuous efforts by the yard's fire brigade and the Kittery Fire Department, the old timber structure was gone is less than an hour, leaving only smoking remains. (NH.)

In a far greater disaster, on May 23, 1939, USS *Squalus* sank during a trial dive, drowning 26 crewmen and shipyard workers in the after compartments. The remaining 33 men in the forward compartments were saved by the use of the new McCann rescue chamber, essentially an improvement on the Momsen diving bell. Here, the final party of survivors, including the commanding officer, Lt. Oliver Naquin, goes ashore from the coast guard cutter *Harriet Lane*. (NH.)

The badly battered *Squalus* was first pumped out, then placed in Dry Dock No. 2 for repair. Note the stabilizing floats attached to her stern on September 15, 1939. Despite heavy damage, she arose nearly brand-new in May 1940—renamed *Sailfish* in an attempt to erase all vestiges of the unlucky *Squalus*—and compiled an impressive record during the course of 12 war patrols. (NH.)

The war behind her, *Sailfish* arrives home to be decommissioned. On her third patrol, early in 1942, *Sailfish* sank the aircraft ferry *Kamogawa Maru*, escorted by four destroyers. On her 10th war patrol, she sank three enemy ships, including the aircraft carrier *Chuyo*, and damaged a fourth vessel. Clearly, *Sailfish* did not inherit *Squalus*'s bad luck. Her conning tower was removed to display at the Portsmouth Navy Yard, and she was scrapped in June 1948. (NH.)

USS *Herring* is launched on January 15, 1942. The new Portsmouth-built boat took part in Operation Torch in November, sinking the merchantman *Ville du Havre* off Casablanca. Thereafter assigned to the Pacific, *Herring* sank six Japanese cargo vessels, but in torpedoing the last two, on June 1, 1944, off Matsuwa Island in the Kuriles, a shore battery registered a direct hit and destroyed her. (NH.)

Photographed during her shakedown trials in the summer of 1943, *Crevalle* became one of the Portsmouth Navy Yard's shining stars. In seven war patrols, she sank nine and a half enemy ships (the half shared with *Flasher*). On her third patrol, *Crevalle* sank the 16,800-ton *Nisshin Maru*, a huge tanker converted from a whale factory. On her fifth patrol, an accident cost the life of the watch officer, Lt. Howard Blind, who had released a critical hatch cover to save the boat. (NH.)

The submarines *Bang* and *Pilotfish* complete construction in Dry Dock No. 2 in the fall of 1943. Between March 1944 and May 1945, *Bang* completed six war patrols and sank eight Japanese vessels. She was converted to a GUPPY boat in 1952. *Pilotfish* made six war patrols between May 1944 and August 1945, performing valuable lifeguard service for downed flyers. She sank no ships, and was expended in a nuclear bomb test in October 1948. (NH.)

With the fall of France and the establishment of the new government at Vichy, the Free French submarine-cruiser *Surcouf*, commissioned seven years before, came to the Portsmouth Navy Yard for a refit. She displaced more than 4,000 tons submerged, carried an aircraft, and mounted two enormous 8-inch guns. She is shown in Dry Dock No. 2 in September 1941. *Surcouf* was later lost in the Caribbean Sea, under suspicious circumstances, in February 1942. (NH.)

Appearing resigned to their surrender, the officers and crew of *U-805* are photographed at the yard before being transported to the naval prison. In addition to *U-805*, three other German submarines were taken to Portsmouth: *U-234*, *U-873*, and *U-1228*. Postwar investigations into irregularities in the prisoners' treatment, including the theft of personal property and the slapping of one U-boat captain, resulted in strong disciplinary measures against senior naval and marine officers. (NH.)

A fine wartime tradition was the donation by the local lodge of B'nai B'rith of kits to the crew members of submarines launched or repaired at the yard. They contained useful personal items and tasty snacks that were unavailable on submarines unless stowed before sailing. Wartime naval officers and B'nai B'rith members pose in Portsmouth with boxes of submariners' kits. Some of these boxes are labeled for USS *Runner*, a submarine that later was lost with all hands on her third war patrol. (PL.)

In an unprecedented dual role, Adm. Ernest J. King served as commander of the U.S. fleet and chief of naval operations during World War II. Highly intelligent and analytical, King graduated first in the U.S. Naval Academy's Class of 1901. After staff service in World War I, he became a submariner and then a naval aviator, both skills valuable in the coming war. King was driven and tactless, and thus widely disliked. He died, at age 77, at the Portsmouth Naval Hospital in 1956. (GW.)

Six

The Last Round

World War II Defenses

Following World War I, the American military establishment entered a period of long decline. The lack of any immediate threat and the diversion of government resources to combat the Depression combined to limit the size and modernization of the military. Seacoast defense was maintained as an active branch during this time, but in a much-reduced status. A very few modern batteries were built in the 1917–1922 period, but then work came to an almost complete halt for more than 15 years. The Portsmouth forts received no new batteries, and just the barest of modernization to the mining and electrical facilities. Additionally, during the war, guns had been taken from two of the larger batteries, leaving Portsmouth with just a minimum of protection.

After the start of World War II in Europe, the nation suddenly awoke to the prospect of its involvement in war again. The United States began a major rearmament program after the fall of France in the last half of 1940. A new series of heavily protected batteries with the most powerful guns was planned for virtually every American domestic port and most offshore bases. As the navy yard in Kittery expanded and played an important part in new submarine construction, so too grew the defenses for Portsmouth. A major new coast artillery post was established at Frost Point in Rye, New Hampshire, and a temporary battery for four 155-millimeter guns was placed on Odiornes Point, pending the construction of the new emplacements. Additional new gun batteries, modernized older batteries and mining facilities, and advanced fire control and communications facilities were added to the older posts. A modern Harbor Defense Command Post (HDCP) and Harbor Entrance Control Post (HECP) were created to coordinate joint army-navy harbor command, and an antisubmarine net was strung across the harbor entrance.

The new 22nd Coast Artillery Regiment was organized just for Portsmouth. While the years between 1940 and 1942 were full of the bustle of new construction and expanding units, the tide of war soon turned against the Axis powers, and the military emphasis shifted from American to foreign theaters. The years from 1943 to 1945 saw a shift of resources elsewhere and, by the end of the war in mid-1945, the coast artillery posts were again neglected sites. In just a few subsequent years, the guns were gone forever and the men discharged to pursue civilian careers and raise families.

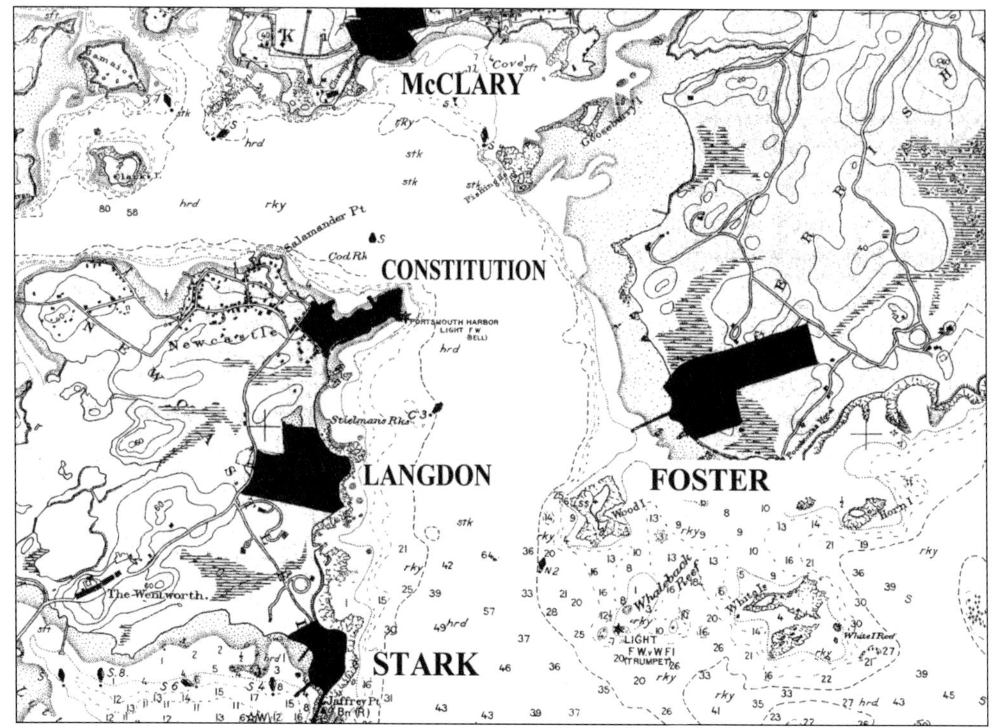

With war looming, one problem facing the army command in Portsmouth was accommodation space. In the late 1930s, four reservations were available. Fort Constitution and Fort Stark held most of the seacoast armament, but were very small. Fort Foster had more acreage, but was not suitable for large garrisons. The Camp Langdon reservation was the only one adequate for new garrison units. This late-1920s map shows the relative position of the sites, as well as Fort McClary, which had reverted to park status by that time. (NA.)

The garrison post located between Fort Constitution and Fort Stark was at first simply named the New Castle Military Reservation. In 1940, it was renamed Camp Langdon. Not emplacing any fixed armament, it was used to house coast artillery units and had adequate room for field training. Over time, the post acquired all the structures of a major army fort. This is Building No. 123, the post signal station in the early 1940s. (NA.)

Along with barracks, quarters, and administration buildings, all the perceived needs to fulfill the troops' daily regimen were required at a post. That included recreational facilities. Camp Langdon's post recreation building, or Building No. RB-1, appears here. It was completed on March 5, 1941, as part of the substantial new cantonment built at the post. The building had a capacity of 364 men, spread over a variety of rooms and facilities. (NA.)

Fire stations needed to be housed on the self-sufficient military reservations. At Camp Langdon, one of the few older buildings retained for use in the World War II period was its firehouse. The building, constructed in 1919, had a garage capacity for four fire trucks. This view shows off one of its machines, parked in front of the station c. 1941. (NA.)

The army eventually made major efforts to improve living conditions at Fort Foster in the early 1940s. Most of the older post buildings were by then in dilapidated condition and were removed. An entirely new cantonment for troops was constructed. This new mess hall, completed on February 17, 1941, occupied 2,605 square feet and had the capacity to serve 250 men. (NA.)

Here, members of the 22nd Coast Artillery stroll around Fort Stark in a seemingly tranquil scene. A dramatic difference in life transpired at the immediate entry of the United States into direct wartime status. This photograph was taken on July 11, 1941. Five months later, the United States was at war, and its military was at the highest level of preparation. (NA.)

The established army reservations were obvious locations to conduct training. A variety of units passed through their gates during the course of the war. Of course, even in the coast artillery, high wartime turnover of personnel meant that drill and training were continuous. In this 1941 photograph, members of A Battery of the 22nd Coast Artillery are framed by old Fort Constitution's gate. The war-duration wooden barracks appear on the right. (PL.)

Inspections of troops and facilities were conducted at least weekly, usually on Saturday mornings, to ensure that military standards of cleanliness and proper uniforms were maintained. They also enforced discipline by making all soldiers conform to orders and made officers aware and thus responsible for their men. At Fort Stark, in 1941, A Battery of the 22nd Coast Artillery stands for rifle inspection before its barracks. (PL.)

As part of their rifle inspection, the men have removed the bolts from their M1903 rifles for closer examination. Lt. Col. Henry J. Cassard, 9th Coast Artillery, asks a soldier to present his bolt. Dirt or powder residue could cause a misfire in combat, and clean bolt and barrel show a well-kept weapon. Fort Stark's World War II cantonment area has recently been expanded, as the shale underfoot and the tarpaper-covered temporary structure in the background indicate. (PL.)

Regular "white glove" inspections kept the barracks clean and orderly and ensured that new soldiers were reminded of the army's chain of command, of which he was at the very bottom. The wooden walls of this building have not yet been painted, indicating it was completed in the early 1940s, before the decision to enhance morale by improving the appearance of soldiers' quarters. (PL.)

This morning's inspection includes the mess hall and kitchen. As an army marches on its stomach, both morale and health depended on the cooks' skills and cleanliness. After the first man to see the officers called "Attention," an officer quickly responded with "Carry on," because the cooks, including the dour one at the coal stove, could not interrupt their lunch preparation. The KP in the doorway, a soldier assigned to duty as "Kitchen Police," assisting the cooks and cleaning the mess hall, has snapped to attention. (PL.)

Fort Foster's battery band practices outside of its barracks during World War II. Army bands were important in providing spirit in parades and marches, often accompanying troops during basic training to enliven long, forced marches. "Jody was there when you left. You're right!" and other ditties, some rather off-color, sung to count cadence, were popular with drill sergeants, but seldom with the tired troops. (KM.)

A special occasion at a Camp Langdon World War II mess hall finds beer on the tables. Battery messes, which served only 120 men of one coast artillery battery, were more popular than consolidated messes that served 1,000. A battery's cooks, living among the men they fed, were able to scrounge fresh eggs and fruit, whereas consolidated messes used powdered eggs and canned peaches. Battery cooks could even make chipped beef on toast, unfondly known as "something or other on a shingle," more palatable. (NC.)

Men write letters and read at Camp Langdon's post library during World War II. The Class A uniform for enlisted men included low quarter shoes and a necktie tucked into the shirtfront. The buck sergeant on the right has a combat rank, while the two men writing are technical specialists, indicated by the "T" under their chevrons. Though the pay was the same, "techs" did not normally command men. (NC.)

The artillery engineer's staff poses in the drafting room at Camp Langdon during the war. The artillery engineer of a harbor defense had the responsibility for the operation, maintenance, and repair of the fire control, communications, and searchlight systems, and of their powerplants and power lines. The master gunner reported to the artillery engineer and often there was an overlap in the membership of their staffs. (NC.)

Soldiers spend part of their day in Camp Langdon's rifle range in this 1940s photograph. The frames upon which canvas targets ride up and down are on the right, while the overhang keeps sand and pebbles from raining down when one of their buddies up-range aims too low. One of these soldiers holds a long stick with a metal disc on the end to show the location of hits; another soldier, on the right, has his hand on a pole with "Maggie's drawers," a red flag, to indicate a complete miss. (NC.)

Many light 3-inch and 6-inch gun batteries were retained in the modernization plan to provide continued coverage of the harbor's minefields. Here, one of the older Model 1902 3-inch gun batteries, likely Battery Hackleman at Fort Constitution, is exercised by its gun crew. The modern helmets indicate that this must date from later in the war, probably 1943. Note the camouflage net spread out overhead to hide the battery from aerial spotting. (NC.)

During the early period of the war, active training on all methods of defense was intensely practiced. This c. 1943 photograph was likely taken at Camp Langdon. Being used here are four of the most important army light arms—the M1 Garand rifle, the M1903 Springfield rifle, the Browing automatic rifle, and a .30-caliber water-cooled machine gun, which has been named "Gypsy Rose Lee," possibly because its barrel became as hot as the famous burlesque stripper. (NC.)

A powerful new mine was developed for army service during the war. Intended for placement on the sea floor, this large device could carry 1,000 pounds of TNT explosive and be detonated both by influence and direct control. Adopted in late 1942, it soon replaced older mines in most American harbors. This photograph shows members of the Portsmouth mine company pushing one of these huge devices on its trolley along the rails. Note, in typical army fashion, how seven men are watching five do the work. (NC.)

To plant the minefield, a small flotilla of boats was necessary. This is a view of the Fort Constitution mine wharf during World War II. The small craft are distribution box boats, also known as L-boats, which are crewed by army, not navy, personnel. They planted the electrical junction boxes that joined the mines to their electrical cables. (NC.)

The loaded mines themselves were usually carried out to the minefields and deposited by special mine planters. *General Absalom Baird*, the army mine planter shared during the 1930s and 1940s by all New England harbor defenses, was built at Milwaukee in 1919. At 172 feet in length, she displaced 704 tons. After the war, she was sold into private hands and entered harbor boat service in New York City. (NC.)

Immediately prior to World War II, the Fort Constitution mining casemate was extensively modernized. A new exterior and entrance were added to the structure. The essential elements of the defenses—new gun batteries, command bunkers, powerplants, and mining casemates—were provided with protective equipment, filters, and seals to prevent incapacitation from poisonous gas. This aerial photograph shows the rebuilt exterior of the Fort Constitution casemate, and in the upper left, the new minefield fire control station built atop Battery Farnsworth. (GW.)

In 1942, soldiers of the Fort Stark meteorological section obtain data on wind velocity and direction by releasing a helium-filled weather balloon and tracing its path with a theodolite. Because the flight of large-caliber shells fired from seacoast forts would be affected by temperature, barometric pressure, and wind, current data were crucial in hitting an enemy warship. Every fort had its complement of trained meteorologists and the necessary equipment. (NC.)

A naval cable barge operates in Portsmouth Harbor during World War II. Such barges were used to deploy the harbor antisubmarine net. Also, both the army and navy depended on communication cables to connect their dispersed bases. The army used insulated electric cables to connect each of its submarine mines to its respective mine casemate, as well as to connect the numerous and distant fire control stations to the command post. (NC.)

An example of a Portsmouth fire control tower, this eight-story silo was located just south of Fort Dearborn on the separate, small reservation at Pulpit Rock. The tower, built between 1942 and 1943, served with two observing stations, one each for the 16-inch and 6-inch gun batteries at Fort Dearborn. On the rooftop there was also an antiaircraft spotting station. The crews for these stations were usually quartered in a small barrack located at or near the base of the tower. Fourteen fire control stations existed for Portsmouth. (GW.)

To obtain the necessary visibility, fire control stations had to be elevated, which also meant they were more conspicuous to an enemy. Consequently, attempts were made to disguise some structures as typical beachside property. The station (upper left) at Bald Head Cliff in York, Maine, represents a standard cottage design. The illustration is from a 1960s postcard for the Cliff House resort. The building and tower were subsequently torn down. (GW.)

After the fall of France to Germany in 1940, American planners accelerated the construction of an entirely new generation of defenses. The primary new weapons were to be powerful 16-inch rifles mounted in heavily protected concrete emplacements. Portsmouth was to receive one of these dual emplacements, at a new reservation on the south flank of the harbor in Rye, New Hampshire, begun simply as Battery Construction No. 103. This is one of a series of construction photographs taken in September 1942. (NA.)

The construction technique used for these large emplacements was termed "cut and cover." Essentially, the ground was leveled, the concrete of the battery was poured, and then a heavy layer of earth was placed on top. Here, the construction of the central corridor is well under way. The guns were placed 500 feet apart and were connected by a protected corridor containing separate powder and projectile rooms, along with a dedicated powerplant and tool, store, and equipment rooms. (NA.)

The guns themselves were housed in casemated emplacements at each end of the corridor. In this September 1942 photograph, the front wing walls of the gun room take shape. The only exposed part would be the opening directly in front, and there a 4-inch steel shield would surround the gun to provide protection from splinters generated by near misses. While not literally impregnable, these types of emplacements could withstand punishment to a level never previously seen in American coastal defenses. (NA.)

This aerial photograph of the Fort Dearborn reservation in Rye, New Hampshire, dates from 1951. Close examination reveals the trace of Battery Seaman (two 16-inch guns) just inland (to the left) of Frost Point, the point of land with a breakwater at upper left of center. Shortly after World War II, the property was managed by the U.S. Navy, but eventually became Odiorne Point State Park. (NO.)

At Fort Foster on the Maine shore, a modern 6-inch gun battery was constructed to complement the heavy fire from Fort Dearborn. Battery Construction No. 205 was located on a new addition to the east of the original plot. Work was not begun until later in the war, and while structurally complete, the battery never received its armament or much of its internal equipment. This recent aerial view clearly shows the two circular concrete aprons around each gun position. (GW.)

The other of the two modern 6-inch gun batteries constructed for the harbor defenses of Portsmouth, as a result of the 1940 program, was completed. These weapons were housed in 4-inch cast-steel wraparound shields, and were placed in the open, outside of their magazines. Battery Construction No. 204 was located at Odiornes Point at the Fort Dearborn reservation. The photograph was taken in 1944, as the gun crew prepares to proof-fire its weapons. (NC.)

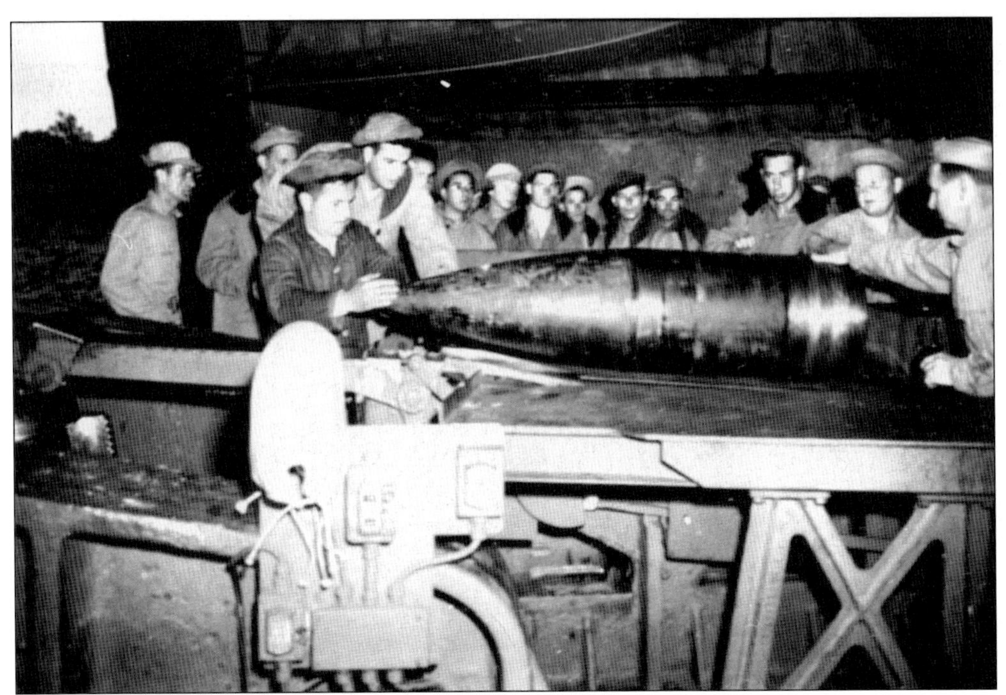

The new heavy battery plan placed the guns in concrete- and earth-covered casemates. While this restricted somewhat the potential field of fire, the gun was very well protected from return fire. Here, the crew of Battery Seaman practices loading huge 2,240-pound projectiles into the breech of one of the guns. The design meant they were inside a casemated gun emplacement and safe from enemy fire. (NC.)

Battery Seaman at Fort Dearborn was built between April 1942 and August 1944. The battery was armed with two ex-naval 16-inch Mk II guns that could elevate 47 degrees for a maximum range of more than 45,000 yards. This photograph is easy to date, as the battery only fired once in its entire life. At completion, the guns were proof-fired in June 1944. Almost immediately, the battery went into reduced-readiness status. (NC.)

As World War II progressed in the Allies' favor, the military became more convinced that the East Coast was not in danger from attack by enemy warships. Moreover, the threat from submarines putting commando parties ashore or attacking shipyards and docks could be met by coast guard patrols and small-caliber, rapid-fire guns at harbor entrances, manned by a reduced number of soldiers. Many coast artillerymen were, therefore, reassigned to antiaircraft units overseas or retrained as infantry to replace casualties. Above, the men of the 22nd Coast Artillery at Camp Langdon pose in 1944 for a last photograph at their old post; they will soon head for Camp Perry, Texas, to earn their combat infantry badges. Portsmouth's rail depot, below, finds these soldiers and their buddies in formation, awaiting their names to be called before boarding. Veterans returning to Portsmouth told of how their new drill instructors in Texas, all experienced in overseas combat, were especially hard on them because of their previously safe and soft life stateside in the coast artillery. (NC.)

In January 1945, Camp Langdon was buried under a typical New Hampshire snowstorm, presenting soldiers with problems civilians are also familiar with. Above, an officer and two enlisted men have trudged out of the headquarters battery orderly room on a recently shoveled path, smiling while one is about to pelt his photographer buddy with a snowball. Below, a truck is assisted out of a snowbank by an Engineer Corps bulldozer. While such tracked equipment could be used to plow roads, soldiers were detailed to clear the narrower sidewalks and were often employed to shovel roadways, either because plows were busy elsewhere or because their commander decided such exercise would keep them occupied when outdoor military training was impossible. Happily, mess hall cooks often decided then, too, to make extra hot chocolate and truck it out to the troops who were freezing on a snowbound road. (NC.)

The 4,255 acres of Pease Air Force Base began as the 300-acre Portsmouth Airport in the 1930s, becoming a military facility during World War II, when the navy leased the property. Transferred to the air force in 1951, the airport was expanded into a Strategic Air Command base for long-range B-47 and B-52 bombers. The base was named for Capt. Harl Pease of Plymouth, New Hampshire, a B-17 pilot who was posthumously awarded the Medal of Honor in 1942 for heroism against the Japanese. (NA.)

A Boeing B-47 Stratojet strategic bomber soars above New Hampshire in the 1950s, heading for a mission near the Soviet Union. Armed with atomic bombs, this six-engine jet had a range of some 3,500 miles and could be refueled in flight, making missions as long as 36 hours possible, although very tiring for its three-man crew. After 14 years of service, these bombers were replaced with faster B-52s that had an un-refueled range of more than 8,500 miles. (NA.)

An unidentified naval tug, with a quite identifiable logo on the stack, goes about her business. Perhaps because of sexy wartime art on combat aircraft, that most tradition-bound service, the U.S. Navy, was finally able to relax and smile a little in the decades afterward. Water cannon above and forward of the pilothouse permit naval tugs to serve equally well as fireboats. (NH.)

The worry and sacrifices of so many civilians during four long years of war helped bring about this happy August 1945 night in Portsmouth. The celebration of VJ Day, which began that morning, continued for more than another week. Every town and city across the country experienced this joy and relief. Many service personnel would not return alive from combat in Africa, Europe, and the Pacific, but many more Allies would have died had not two atomic bombs convinced the Japanese to end the war they had started. (PL.)